BARRON'S

NEW JERSEY

GRADE 4

MATH TEST

Luann Voza, Ed.D.

All inquiries should be addressed to:
Barron's Educational Series, Inc.
250 Wireless Boulevard
Hauppauge, NY 11788
www.barronseduc.com

ISBN: 978-1-4380-0728-1

Library of Congress Control Number: 2015946199

Date of Manufacture: August 2015
Manufactured By: B11R11

Printed in the United States of America
9 8 7 6 5 4 3 2 1

Contents

Introduction to the PARCC

CHAPTER 1

IMPORTANT NOTE

Barron's has made every effort to ensure the content of this book is accurate at press time. However, the PARCC Assessments are constantly reviewed and revised. Be sure to check *www.parcconline.org* for all of the latest testing information. Regardless of the changes that may be announced after press time, this book will still provide a strong framework for fourth-grade students preparing for the assessment.

For Students

All fourth-grade students in New Jersey take an important test in the spring. This test is called the PARCC Assessment (Partnership for Assessment of Readiness for College and Careers). The PARCC Assessment has two different tests that cover mathematics and language arts.

Not only are there two tests, but both tests will be taken on the computer. That may be enough to make your knees shake—YIKES! Believe it or not, you really do not need to be nervous. You and your teachers have been preparing for this since you began kindergarten.

Just in case those words of assurance do not completely convince you, this book is designed to prove to you that YOU ARE READY. The helpful hints and sample tests included here will give you the confidence you need to do your very best work on the mathematics section of the tests. These have been tested in a classroom with other fourth-grade students just like you, and they work.

Here is some information about each of the tests that will help you understand what you will need to do.

The PARCC is given after about 75 percent of the school year is completed. The test contains different types of items, such as multiple choice, short answer, computational, and extended types of items that may involve how well you understand math concepts and how well you can apply the math skills learned in class. In addition, you may be asked to give explanations or create models to show how you found your answer. The PARCC test includes three different items, with point values between 1 and 6. Some of the items will be hand scored and some machine scored.

For the assessment, you will use a computer to complete the tasks. You will use computer-based tasks such as:

- Select (multiple-choice)
- Select (multiple-choice with more than one correct answer)
- In-line choices (using a drop-down menu)
- Drag and drop
- Fill-in-the-blank (by typing in your answer)
- Combination equation builder and text editor (to type in both an equation in numbers and an explanation in words)
- Hot spots (provide correct answers by selecting an object such as a point on a number line)

Practice your typing skills so that you can easily type your responses. Also practice using the mouse or touch pad/touch screen to select items, drag and drop, highlight, scroll, and type your answer (whole numbers and fractions) in the box shown. By knowing what to expect and practicing your skills in advance, you will give yourself the best possible chance to succeed on both parts of the PARCC Assessment in mathematics.

In this book, you will find each of the main sections of the test explained in its own chapter. Each chapter contains additional opportunities for independent practice and a detailed explanation of the correct answers. Each chapter also has a checklist for you to use to make sure that you have reviewed all of the concepts and skills covered in that chapter. In addition, there are two complete practice tests at the end of the book.

Overview for Parents and Teachers

This book is written to help New Jersey fourth-grade students achieve proficiency on the mathematics portion of the PARCC Assessment. In addition to the practice tests, it includes instructional study units in the following skill areas:

1. Understanding test-taking strategies
2. Solving multistep complex problems with work or explanations of solutions

3. Expressing reasoning and constructing arguments to demonstrate conceptual understanding
4. Practicing computation to improve fluency
5. Utilizing appropriate mathematical tools strategically
6. Constructing and utilizing mathematical models and strategies

What to Expect

Students will be introduced to strategies that will help build familiarity with the test format while gaining confidence in the individual tasks required. Checklists are provided and explained to help inform students of what is expected and how they will be scored. Ample opportunities will be available for both guided and independent practice of each portion of the test.

Testing Times and Formats

The PARCC Assessment takes place over a single, 30-day testing window for computer-based tests. The test will consist of four different units, each with a testing time of 60 minutes for a total testing time of four hours.

Overview of PARCC Mathematics Task Types

Type I: Tasks assessing concepts, skills, and procedures

These items involve a balance of computation and explaining how you found your answer. These items will be machine scored and are worth 1–2 points each.

Type II: Tasks assessing expressing mathematical reasoning

These items may ask you to explain how to solve a problem, or decide if a given solution is correct or incorrect. These items will be machine scored and are worth 3–4 points each.

Type III: Tasks assessing modeling and applications

These items may ask you to create a model, such as an area model, equation, or number line, to solve a problem that is about a real-world situation. These items will be hand scored and are worth 3–6 points each.

How the PARCC Is Given

The fourth-grade PARCC Assessment is taken on the computer using an online testing system called TestNav. Students will be asked to read items and answer different types of questions on the computer. Examples of the different types of questions can be found at *http://www.parcconline.com*.

The Common Core State Standards

Notes to Parents and Guardians

The material covered on the fourth-grade PARCC Assessment is designed to measure students' achievement based on the Common Core State Standards (see Appendix).

These standards were developed to provide a consistent, clear understanding of what students are expected to learn at each grade level. These standards were adopted by forty-five states and provide guidelines so that teachers and parents know how to help prepare students for college and career.

The test is cumulative in that these skills are not taught in isolation but, rather, are built upon each school year.

Your child will be expected to answer various types of questions. The chapters in this book provide examples of these types of questions.

Some students are unnerved by the fact that the test is timed. They worry that time will run out before they finish. Practice tests like the ones included in this book are a good way to put these fears to rest. As the skill becomes more familiar, it becomes less frightening.

In preparing for this test, common sense should prevail. Provide a quiet place to complete homework every evening. Encourage daily reading and time for both parent and child to read aloud. Talk about school experiences and get a sense of how your child feels about what he or she is learning in fourth grade.

In addition, because the test is computer based, students should practice typing skills and be able to work comfortably on whatever testing device their school district plans to use during the assessment period (PC, laptop, tablet device, etc.).

As with any standardized test experience, a good night's rest and healthy breakfast provide a testing advantage. Feeling good physically can promote a positive attitude toward the test experience.

Finally, after all the preparation is done and the tests are over, the most common question is usually "When can I expect to see results?" School districts typically receive test scores sometime during the months of June–August. However, it is not uncommon for scores to be distributed early in the following school year.

What Is the Common Core?

The Common Core refers to a common and clear set of guidelines about what all students should know in math and language arts in grades K–12. These guidelines are designed to help prepare students for college and/or career.

What Are the Common Core Guidelines?

The standards provide guidelines for teachers and parents about what should be taught at each grade level. Teachers still have flexibility in how the content is taught. These rigorous standards will prepare students for the twenty-first century. Forty-five states, as well as the District of Columbia, are using these standards.

How Does This Impact Education?

New Jersey has had standards for many years. The Common Core is the newest set of standards adopted by New Jersey. Standards set a goal that students must meet by the end of the school year. The standards do not tell teachers how to teach, but they tell teachers what content must be covered in a specific grade level. The PARCC test was developed to determine how successful students are with mastering these standards.

Information About the Standards

The standards are broken into several sections:

- Operations and Algebraic Thinking
- Numbers and Operations in Base Ten
- Numbers and Operations—Fractions
- Measurement and Data
- Geometry

Each section outlines some of the things that students should be able to do in each area, and at each grade level. A complete list of the standards can be found in the Appendix.

Operations and Algebraic Thinking

This standard focuses on the four mathematical operations (addition, subtraction, multiplication, and division) with whole numbers and how they are used to solve problems. Students should be able to construct and solve equations that involve mathematical and real-world applications. In addition, students should be able to assess the reasonableness of their solutions using estimation strategies and explanations.

Numbers and Operations in Base Ten

This standard focuses on place value and the formation of multidigit whole numbers. Understanding of place value is used to read, write, compare, and order whole numbers. Computational fluency in all mathematical operations involving whole numbers is accomplished with algorithms and models.

Numbers and Operations—Fractions

This standard focuses on fraction equivalence, comparisons, and operations through the use of number lines and other visual fraction models. Decimal notation is included with an emphasis on reasoning through the use of benchmarks and other visual models.

Measurement and Data; Geometry

This standard focuses on problem solving involving measurement and conversions of units within a single system of measurement. Informational data is represented and interpreted through displays such as line plots. Geometric measurement and conceptual understanding of angles is included along with classification of two-dimensional shapes based on properties.

Standards for Mathematical Practice/ Problem-Solving Strategies

CHAPTER 2

In this chapter, you will focus on problem solving. It is helpful to think of word problems, not as problems, but as situations that you may face in the real world that require some mathematical skills to find or solve an unknown part of the situation. Like anything, practicing a skill will always make you better at the skill. Successful math students are good at sticking with a problem and finding ways to solve problems even if their first solution is not correct.

The Common Core State Standards for Math are based on eight Mathematical Practices (MP), or traits, that students should have in order to be successful in solving problems.

Mathematical Practices (MP)

1. **Make sense of problems and persevere in solving them.** Make a plan and carry it out. Keep trying until a solution can be found. Do not give up even if your first attempt is not a success.
2. **Reason abstractively and quantitatively.** Explain how you got your answer.
3. **Construct viable arguments and critique the reasoning of others.** Decide if the solutions of others are correct and explain why or why not.
4. **Model with mathematics.** Use visual models such as number lines, arrays, and area models to solve problems.
5. **Use appropriate tools strategically.** Decide which tools, such as rulers and protractors, are useful and use them correctly.
6. **Attend to precision.** Check your work for errors in computation, units of measurement, and vocabulary. Use correct symbols and forms of numbers.
7. **Look for and make use of structure.** Use what you have learned so far to solve problems.
8. **Look for and express regularity in repeated reasoning.** Look for patterns or rules that are used to help find shortcuts for solving problems.

Steps for Problem-Solving Success

Math students who can easily solve problems either have a natural math instinct for numbers or, more likely, have found a set of steps that are helpful. Here are some steps that you can use.

1. **Read the problem.** Look for key words or phrases that can guide your thinking. Try to restate the problem in your own words. Problems might remind you of a similar problem that you may have solved before.
2. **Think about what you know and what you don't know about the problem.** Organize your thoughts and try to write an equation that may be used to show the known values and either a letter or a question mark for the unknown values. For more complex problems, more than one equation may be needed.
3. **Create a visual model that can be used to represent the problem.** Models such as number lines, arrays, shapes with fractional or decimal shaded amounts, or geometric figures are examples of visual models. Others may include bar models, place value charts, and tables or graphs.
4. **Decide on a solution by working out the problem and solving the equation that has been created.** Check your solution by substituting values that are now known to see if they may be correct. Check all calculations for accuracy.
5. **If the solution appears not to be correct, try to revise any part of the equation.** Think about a different way to approach the problem such as acting it out or drawing a picture. The key is to keep trying, maybe by trying a different equation or operation.

Problem-Solving Strategies

1. **Draw pictures.** When you can see the math, you are more likely to understand what is known and unknown about the problem. Pictures can include equations, lists, charts, graphs, bar models, figures, or boxes with numbers in them.
2. **Make the problem simpler to understand.** Try substituting smaller values that may be easier to work with first to find a strategy. If successful, then replace the values with the ones from the problem.
3. **Work backward to solve.** If the total is known, and you need to find out how it was found, write the equation with the unknown value and work backward from the end, in many cases using the inverse, or opposite, operation.
4. **Guess and check.** Make a first attempt at solving the problem. If it does not appear to be correct, work through the problem to see if there was a slight error in calculation, incorrect operation, or start over with a new strategy.

Examples

1. Groups of students are collecting plastic bottles for a science project. They set a goal of collecting a total of 500 bottles.

   Alana's group collected 145 bottles.

   Brendan's group collected 36 more than Alana's group.

   Carlo's group collected 12 less than Brendan's group.

   Did the students meet their goal of collecting 500 bottles? Why or why not?

Restate: Does $A + B + C = 500$ or more?

Known values: Alana's group (A) = 145

Unknown values: Brendan's group (B), Carlo's group (C), total bottles (T)

Key words: 36 more than, 12 less than

Visual model:

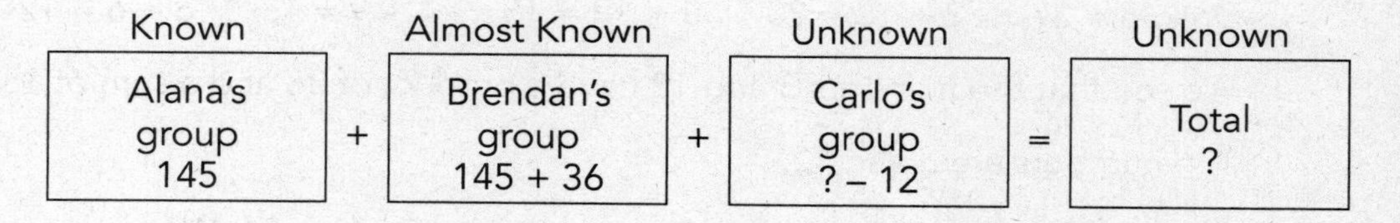

We can now see the problem, what is known/unknown, and how to find the total.

Begin by solving what is almost known:

$$145 + 36 = 181 \text{ (Brendan's group)}$$

Now we can solve for Carlo's group:

$$181 - 12 = 169 \text{ (Carlo's group)}$$

We can now find the total:

$$145 + 181 + 169 = 495$$

We can now answer the question. No, the students did not reach their goal of 500 bottles. They collected a total of 495 bottles:

$$495 < 500$$

The problem was solved by using Steps 1–4 and Strategy 1. MPs 1, 2, 4, and 6 were also used.

2. Mr. Perez wrote a math riddle on the board.

The sum of two numbers is 15.

The product of the same two numbers is 36.

What are the two numbers?

Restate: When $A + B = 15$ and $A \times B = 36$, what do A equal and B equal?

Known: the product and the sum of two numbers

Unknown: the value of the two numbers

This problem can be solved by making a list. Since the list of numbers with a product of 36 is shorter than the list of numbers with the sum of 15, begin with that list.

Factors of 36 (factors are the numbers we multiply to get a product):

$1 \times 36 \quad 2 \times 18 \quad 3 \times 12 \quad 4 \times 9 \quad 6 \times 6$

Now find the sums of the same two numbers:

$1 + 36 = 37 \quad 2 + 18 = 20 \quad 3 + 12 = 15 \quad 4 + 9 = 13 \quad 6 + 6 = 12$

We see that the numbers 3 and 12 have a product of 36 and a sum of 15.

The numbers are 3 and 12.

We used Steps 1, 3, and 4 to solve, and Strategies 1 and 4. We also used MPs 1, 4, and 6.

3. Francesca and some friends ate lunch at her house. After lunch, there were 4 bowls of soup left. The amount of soup in each bowl was $\frac{1}{3}$ of a pint. Francesca's mom wanted to put the remaining soup in 1-pint containers. How many containers would be needed for the remaining soup?

Known: the amount of soup in each of 4 bowls

Unknown: the total amount of soup and whether or not the total is greater than or less than one whole

Begin by drawing a model to show the bowls and the amounts.

In the model create an equation that can be used to solve:

= ?

$\frac{1}{3}$ $\frac{1}{3}$ $\frac{1}{3}$ $\frac{1}{3}$

Find the total by adding the fractions. Since they are like fractions, they can be easily added:

$$\frac{1}{3}+\frac{1}{3}+\frac{1}{3}+\frac{1}{3}=\frac{4}{3}$$

Since the total is more than one whole pint, a 1-pint container will not be enough.

Francesca's mom will need 2 containers for the remaining soup.

This is because the total is $\frac{4}{3}$, or $1\frac{1}{3}$, which is more than 1 pint and less than 2 pints.

We used Steps 2, 3, and 4 to solve, and Strategy 1. MPs 1, 2, 4, 6, and 8 were also used.

4. There were 6 large boxes, each with 175 bananas that were unpacked at the fruit market. After removing 23 bananas, the remaining bananas will be repacked into smaller bags, each with 8 bananas. How many full bags of 8 bananas can be filled with the remaining bananas?

Known: 6 large boxes, each with 175 bananas with 23 removed

Unknown: the number of smaller bags of 8 of the remaining bananas

This problem can be solved by first writing an equation to find the total number of bananas in the boxes:

$$6 \times 175 = ? \qquad 6 \times 175 = 1{,}050$$

Next, take away 23 bananas:

$$1{,}050 - 23 = 1{,}027$$

Then, find the number of bags of 8 that can be made from the remaining 1,027:

$$1{,}027 \div 8 = 128 \text{ R3}$$

Finally, decide what to do with the remainder.

Since the bags must be full, then 128 bags of 8 can be filled.

We used Steps 2 and 4 to solve the problem. We also used Strategy 1 and MPs 1, 2, and 6.

5. The yard at Mr. Athavan's house has a width of 4 meters. The length is 3 times as many meters as the width. Mr. Athavan's son, Frankie, said that the perimeter of the yard is 48 meters. Is Frankie correct? Why or why not?

Known: width of the yard

Almost known: length of the yard

Unknown: perimeter of the yard

This problem can be solved by drawing a picture and labeling what is known:

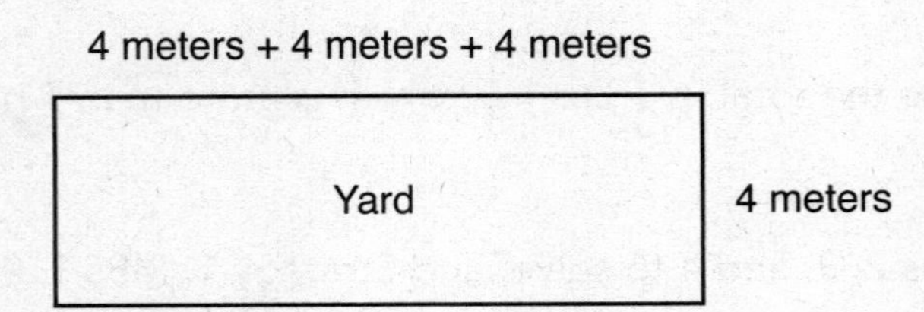

The equation 4 + 4 + 4 = ? can be used to find the length:

$$4 + 4 + 4 = 12$$

The equation 12 + 4 + 12 + 4 = ? can be used to find the perimeter:

$$\text{perimeter} = 12 + 4 + 12 + 4 = 32 \text{ meters}$$

Since 32 meters is the perimeter, then 48 meters cannot be the perimeter.

Frankie is not correct.

We used Steps 3 and 4; Strategy 1; and MPs 3, 4, and 6 to solve this problem.

Problem-Solving Exercises

1. Reanna's dad built an outside deck from pieces of wood. He used 178 pieces for the front, 167 pieces for the side, and 55 pieces for the steps. When he was finished, he had 43 pieces of wood left. How many pieces of wood did Reanna's dad have before he started building the deck?

2. Casey wrote a 6-digit mystery number using the following clues:
   - There is a 2 in the thousands place and in the tens place.
   - The value of the digit in the ones place is 2 times the value of the digit in the thousands place.
   - The value of the digit in the ten thousands place is 3 times the value of the digit in the tens place.
   - The other digits in the number are 0 and 3.

   What is Casey's mystery number? Explain how you found your answer.

3. Kirsten has $10 to buy enough orange juice to fill a 2-gallon container. The orange juice is only available in 1-quart containers for the price of $1.00 for each quart. Does Kirsten have enough money to buy the total amount of orange juice quarts needed to fill the 2-gallon container? (No tax is charged.)

4. The line plot below shows the number of minutes it took some students to complete a math problem.

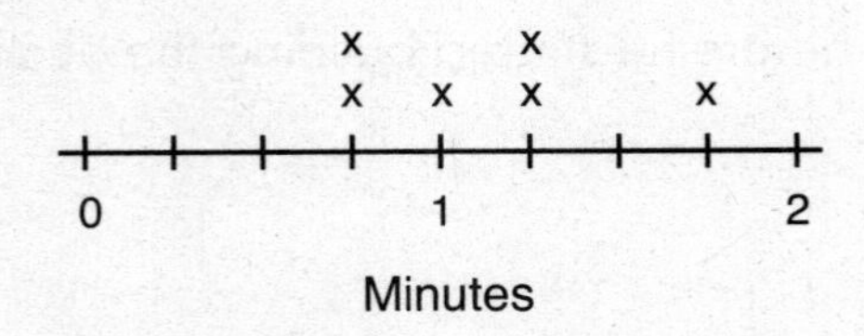

Sebastian also completed the math problem, but his time is not on the line plot. When Sebastian's time is placed on the line plot, the total time for all of the students combined will be 8 minutes. How many minutes did it take Sebastian to complete the math problem? Explain how you found your answer.

5. Lexi shaded the clock below to show how far the minute hand traveled around the clock while she completed her math homework. She wants to measure the angle of the shaded area. How many 90-degree angles make up the total angle of the shaded area on Lexi's clock?

Answers to Problem-Solving Exercises

1. Reanna's dad used a total of 443 pieces of wood. The total amount of wood used plus the number of pieces of wood left over equals the total amount of wood Reanna's dad had in all. A visual model, equation, and working backward are great strategies to solve this problem.

   This bar model can be used to visualize the problem:

Total Pieces of Wood			
178	167	55	43

   This equation can be used:

   total pieces – (178 + 167 + 55) = 43

   Working backward to solve:

   43 + 55 + 167 + 178 = total

   The total number of pieces is 443.

2. Casey's mystery number is 362,024. This can be found by creating a visual model and working from what is known to what is unknown.

   Begin by putting blanks for the 6 digits in the number:

   ___ ___ ___ , ___ ___ ___

   Next, fill in the known numbers:

   ___ ___ 2, ___ 2 ___

   Then, solve for the next clues $2 \times 2 = 4$ (ones place) and $2 \times 3 = 6$ (ten thousands place):

   ___ 6 2, ___ 2 4

   Finally, fill in the last clue. Since a 6-digit number cannot start with a 0, the digit 3 has to be in the hundred thousands place, leaving the digit 0 for the hundreds place:

   362,024

3. Kirsten does have enough money to buy the orange juice. A visual model and an equation can be used to solve the problem:

You need to know that 4 quarts equal 1 gallon. There are 8 quarts in 2 gallons. If each quart has a cost of $1, then 8 quarts have a cost of $8.

Since $10 > $8, Kirsten's amount of $10 is enough.

4. Sebastian's time to complete the math problem is $1\frac{1}{4}$ minutes.

This can be found by writing an equation to solve the total time for the line plot without Sebastian's time. Each *X* has a value of the fraction it represents on the line plot:

2 *X*s on the $\frac{3}{4} = \frac{3}{4} + \frac{3}{4} = \frac{6}{4} = 1\frac{2}{4}$

1 *X* on the 1

2 *X*s on the $1\frac{1}{4} = 1\frac{1}{4} + 1\frac{1}{4} = 2\frac{2}{4}$

1 *X* on the $1\frac{3}{4}$

Sum of the totals: $1\frac{2}{4} + 1 + 2\frac{2}{4} + 1\frac{3}{4} = 5\frac{7}{4} = 6\frac{3}{4}$

To find Sebastian's time subtract the total from 8:

$$8 - 6\frac{3}{4} = 7\frac{4}{4} - 6\frac{3}{4} = 1\frac{1}{4}$$

5. A total of 3 angles, each measuring 90 degrees, can make up the total angle measurement that the minute hand traveled around the shaded area of the clock. This problem can be solved by drawing a model of the clock and placing 90-degree angles on the clock. A 90-degree angle is a right angle.

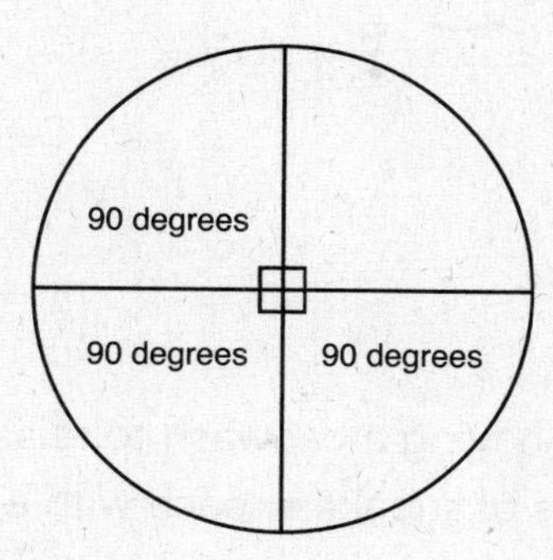

A total of three 90° angles can fit on the shaded area of the clock to show how many degrees the minute hand traveled around the clock in the shaded area.

Test Your Skills

1. Cecelia wrote a number pattern with a rule of "add 3." The fifth number in the pattern is 22. What is the first number in Cecelia's pattern?

2. Tommy's soccer team is having a car wash to raise money to buy sneakers. They want to buy 9 pairs of sneakers, each with a cost of $40. The car wash tickets have a cost of $5. How many car wash tickets does Tommy's soccer team need to sell in order to buy the 9 pairs of sneakers?

3. Kaylee wants to put 5 stickers on a page that has a length of 5 inches across. Each sticker has a length of $\frac{7}{8}$ inch. Can Kaylee fit the 5 stickers across the 5-inch page? Why or why not?

4. Miss Jonas is having 8 parent–teacher conferences at her school. Each conference will have a length of 10 minutes. There will be no time in between the 8 conferences. If Miss Jonas starts her conferences at 2:00 P.M., at what time will the 8 conferences end?

5. Some students are making polygons by using straws for each of the sides. There are a total of 72 straws. So far, the students made 4 triangles, 7 trapezoids, and 3 pentagons. They want to make hexagons with the rest of the straws. How many hexagons can be made with the remaining straws using 1 straw for each side?

6. Dante's room has a wall with a length of 4.25 meters. On the wall, there are 2 windows, each with a length of 0.8 meters. There is also a bookcase with a length of 1.08 meters. Dante wants to put a picture on the same wall in the space that isn't covered by a window or bookcase. What is the length of the remaining part of the wall in Dante's room?

7. Mia has a 4-liter bottle of lemonade that is full. She filled 6 glasses with the lemonade, each with 575 milliliters. How much lemonade, in milliliters, is left in Mia's bottle?

8. The model below shows Jimmy's yard behind his house:

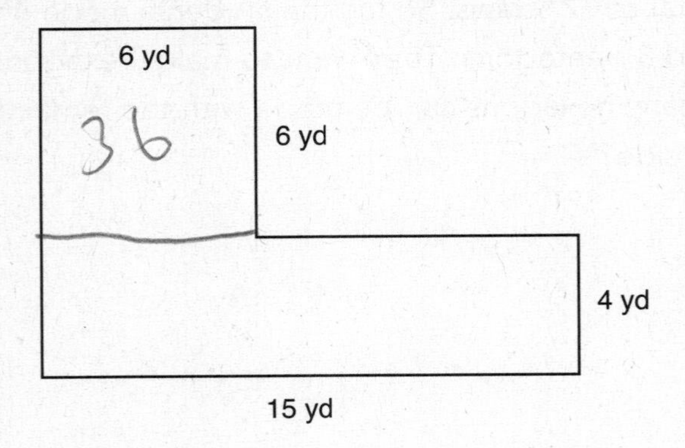

What is the total area, in square yards, of Jimmy's yard?

9. Maria has a fan from her grandmother. She opened the fan to show a total angle measurement of 140 degrees:

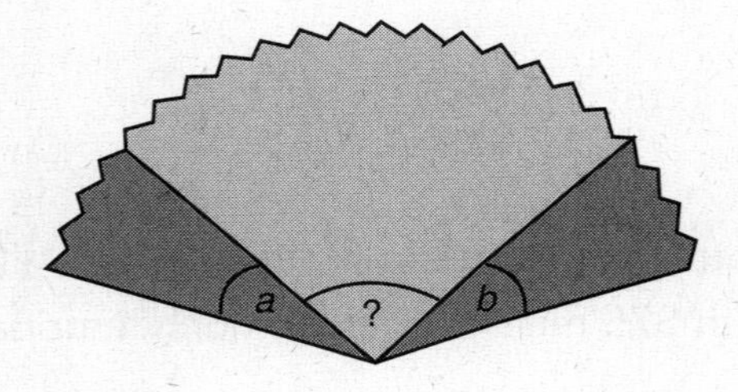

The dark shaded sections of the fan show angles *a* and *b*, each of which has an angle measurement of 15 degrees. The rest of the opened fan is shown by the light gray section. What is the angle measurement of the light gray section of Maria's fan?

Answers to Test Your Skills

1. Number pattern: 10, 13, 16, 19, 22. Starting number is 10.

   The answer can be found by subtracting 3 from 22 four times

   $$22 - 3 = 19,\ 19 - 3 = 16,\ 16 - 3 = 13,\ 13 - 3 = 10.$$

   The first number can also be found by multiplying 4 by 3 to get a product of 12, and then subtracting 22 – 12 = 10.

2. A total of 72 tickets need to be sold.

   9 × 40 = 360; total cost of the sneakers is $360

   360 ÷ 5 = 72; the cost of the sneakers is divided by the cost of one ticket to find the number of tickets needed to raise $360.

3. Yes, the 5 stickers can fit across the page.

   $$5 \times \frac{7}{8} = \frac{35}{8} = 4\frac{3}{8} \text{ inches}$$

   Also, if each sticker is less than 1 inch, then 5 stickers will be less than 5 inches.

4. If the conferences start at 2:00 P.M., then they will end at 3:20 P.M.,

   8 × 10 = 80 minutes, which is the same as 1 hour and 20 minutes.

   Starting at 2:00 P.M. and ending 1 hour and 20 minutes later results in an ending time of 3:20 P.M.

5. A total of 2 hexagons can be made from the remaining straws.

   Triangles: 4 × 3 = 12 straws

   Trapezoids: 7 × 4 = 28 straws

   Pentagons: 3 × 5 = 15 straws

   Total from the 3 shapes: 55 straws

   72 – 55 = 17 remaining straws

   A hexagon has 6 sides.

   17 ÷ 6 = 2 R5

   2 hexagons (6 sides each) = 12 with 5 straws left

6. The remaining wall space is 1.57 meters.

$$0.8 + 0.8 + 1.08 = 2.68$$

$$4.25 - 2.68 = 1.57$$

7. There are 550 milliliters left in the bottle.

   There are 1,000 milliliters in 1 liter, so 4 liters = 4,000 milliliters.

   $6 \times 575 = 3{,}450$ milliliters

   $4{,}000 - 3{,}450 = 550$ milliliters.

8. The total area is 96 square yards.

   Square area: $6 \times 6 = 36$

   Rectangle area: $15 \times 4 = 60$

   $36 + 60 = 96$

9. The measurement of the light gray angle is 110 degrees.

   15 (angle *a*) + 15 (angle *b*) = 30

   $140 - 30 = 110$.

Operations and Algebraic Thinking (OA)

CHAPTER 3

VOCABULARY

Equation: a number sentence used in mathematics that usually involves equal-sized amounts on either side of an equal sign; some of the amounts are known and some may be unknown

Operation: one of four ways to compute or solve a number sentence; we use four different operations (addition, subtraction, multiplication, and division) when solving number sentences

Factors: numbers that are used in multiplication when joining or combining equal amounts or resizing one value based on a second value

Product: the number or amount that represents the total when combining equal groups or the new amount when resizing an original value

Note

We multiply factors to find the product:

factor × factor = product

Example: The numbers 3 and 4 are factors of the number 12 because $3 \times 4 = 12$.

Multiples: a set of numbers that can be evenly grouped by the same number

Example: The first five multiples of the number 4 are 4, 8, 12 ,16, and 20. They are multiples of 4 because they can be evenly divided, or put into equal groups of 4 without having a remainder.

Prime numbers: numbers that have exactly one factor pair (1 and itself)

Composite numbers: numbers that have more than one factor pair

Operations

Multiplication

We work with multiplication when we want to join together equal groups, or resize one amount based on a second amount. We can also use multiplication for models such as an array or area model.

Example 1

There are 5 flowers in the small vase.

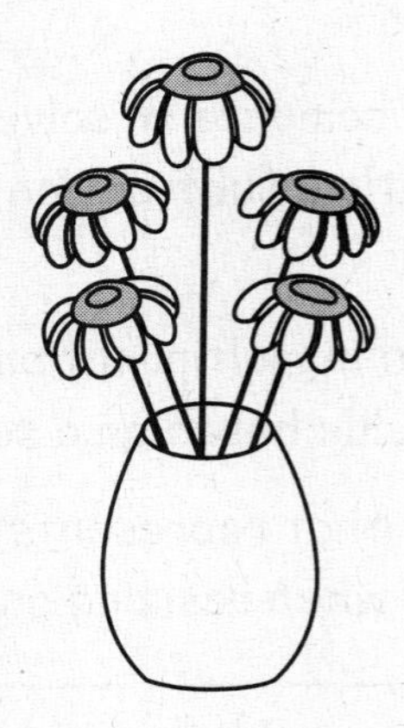

There are 3 times that amount of flowers in a large vase.

We can see that $5 \times 3 = 15$. The number 15 is 3 times as much as 5. It is also 5 times as much as 3. When we resize the number 5, the amount is 3 times as much, which equals 15.

Example 2

The length of a closet is 4 feet.

4 feet

The length of the hallway is 6 times the length of the closet.

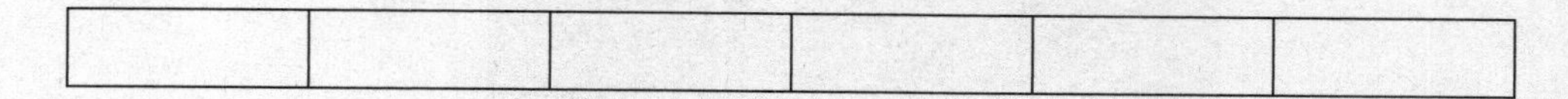

The length of the hallway is 24 feet because $4 \times 6 = 24$.

Example 3

Paige wants to place some stickers on her notebook cover. She wants to put 6 stickers in each of 7 rows as shown:

Question: How many total stickers are on Paige's notebook cover?

We can use the equation $6 \times 7 = ?$ to solve. Since the product is 42, we can solve to find the total of 42.

Answer: Paige has a total of 42 stickers.

Example 4

The bulletin board in Ms. Garcia's classroom has a length of 8 feet and a width of 4 feet.

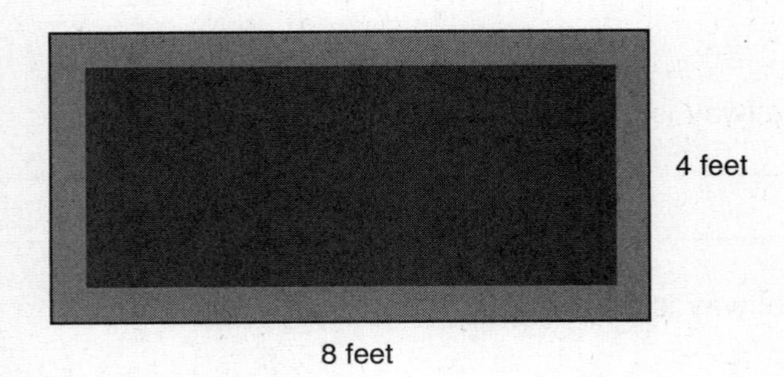

Question: What is the total area, in square feet, of Ms. Garcia's bulletin board?

We can use the equation $8 \times 4 = ?$ to solve. Since the product is 32, we can solve to find the area of 32 square feet.

Answer: The area of Ms. Garcia's bulletin board is 32 square feet. Note that we use the measurement square feet for area because it has two dimensions: length and width.

Division

Division is the grouping of an amount. We can either find the number of equal-sized groups or the amount in each group. We also use division for models such as an array or an area model.

Example 1

Vinnie runs a total distance of 20 miles each week. He runs the same amount of miles each of 5 days.

Question: How many miles does Vinnie run on each of the 5 days?

We can model the problem with the equation $? \times 5 = 20$ miles.

We can also use the equation $20 \div 5 = ?$ to solve.

We know that $20 \div 5 = 4$.

Answer: Vinnie runs 4 miles each day.

We can check our answer with the equation $4 \times 5 = ?$. Since $4 \times 5 = 20$, the answer is correct.

We can use division when we know the total number and want to make an array.

Example 2

An array of 45 buttons has equal-sized rows with 9 in each row.

Question: How many rows of buttons are there?

We can use the equation $9 \times ? = 45$ or the equation $45 \div 9 = ?$ to solve. We know that $45 \div 9 = 5$.

Answer: There are 5 rows of buttons. We can also use the equation $9 \times 5 = 45$ to check our answer.

We can use division when we know the area and want to find the length of one of the sides.

Example 3

A table has an area of 21 square feet. The width is 3 feet.

Question: What is the length?

We can use the equation $21 \div 3 = ?$ to solve.

Answer: Since $21 \div 3 = 7$, the length of the table is 7 feet. We can use the equation $3 \times 7 = 21$ to check.

Division with Remainders

In some division problems, it is not possible to make equal groups. When this happens there is a remainder, or leftover, amount to the equal groups. For some problems, you must solve the problem while deciding what to do with the remaining parts of a group. In some cases, you may round up to the next whole number when another group must be made to include the remainder. In other cases, you may round back to the nearest whole number when the remainder cannot be included. In other cases, you may need to make unequal groups to include the remainder. Think about each problem to determine what to do with the remainder.

Example 1

There are 28 students who are going to sit in the bleachers to watch a baseball game. Each row of bleachers can fit 6 students.

Question: How many rows of bleachers will the students need so that they all can sit and watch the baseball game?

Use the equation $28 \div 6 = ?$ to solve. We know that we can fit 24 students in 4 rows of 6 because $6 \times 4 = 24$, but that leaves 4 remaining students. This means that we must round up to 5 rows because we need to include all 28 students.

Answer: A total of 5 rows of bleachers are needed to fit 28 students.

> **Note**
> We round the quotient up to the next whole number when we need to include the remainder.

Example 2

Mrs. Akin wants to schedule some time to grade her writing tests. It takes 6 minutes to grade each test. Mrs. Akin has a total of 45 minutes to grade some tests.

Question: How many tests can Mrs. Akin finish grading in 45 minutes?

We can use the equation $45 \div 6 = ?$ to solve. We know that $6 \times 7 = 42$, but that leaves 3 remaining minutes. We cannot round up because 3 minutes is not enough to grade a test, so we can only count tests that could be completed.

Answer: Mrs. Akin can finish grading 7 tests in 45 minutes.

Example 3

Drew is playing soccer. There are a total of 50 boys and girls playing soccer and there are 4 teams. Each team must have no less than 10 players.

Question: How many players could be on each of the 4 teams?

We can use the equation $50 \div 4 = ?$ to solve.

We know that $4 \times 12 = 48$, but that leaves 2 remaining players. We cannot round up because we cannot make another team with only 2 remaining players since we need to have at least 10 players on each team. We must include all 50 players on the teams. We can distribute the 2 remaining players by placing 1 on each of 2 teams. The groups will not be even, but this is as even as they can be.

Answer: There will be 2 teams with 12 players and 2 teams with 13 players.

$$12 + 12 + 13 + 13 = 50$$

Solving Multistep Problems Using the Four Operations

Multistep problems involve more than one step and more than one type of operation to solve. The key to multistep problems is to determine what is known and what is unknown, and then decide which operation or operations are needed to solve.

Example 1

Mr. Reynolds is giving out pencils to his students for their PARCC Practice Test. He has 26 students and wants to give each student 2 pencils. He has 5 full boxes of pencils and each box has 12 pencils.

Question 1: Does Mr. Reynolds have enough pencils so that he can give each student 2 pencils?

Question 2: If he doesn't have enough pencils, how many more pencils does he need? If he has more than enough, how many does he have left?

First, write an equation that can be used to find the total number of pencils Mr. Reynolds needs.

The equation $2 \times 26 = ?$ can be used to solve.

We know that $2 \times 26 = 52$, so Mr. Reynolds needs 52 pencils.

Next, write an equation that can be used to find the total number of pencils Mr. Reynolds has. Since he has 5 boxes of pencils, and each box includes 12 pencils, the equation $5 \times 12 = 60$ shows that Mr. Reynolds has 60 pencils.

Next, answer the first question based on what has been solved so far.

Answer 1: Yes, Mr. Reynolds has enough pencils so that he can give each student 2 pencils.

He has 60 pencils, and he only needs 52 pencils.

Finally, find which part of Question 2 could be answered based on the answer to Question 1.

Answer 2: Mr. Reynolds will have 8 pencils left:

$$60 - 52 = 8$$

We used multiplication, multiplication, and subtraction to solve all parts of the question.

Example 2

Ms. Patel is making PARCC review folders for her students. Each folder needs to include 9 pages of worksheets. She has 2 packs of paper to use. The first pack has 85 sheets of paper and the second pack has 33 more sheets of paper than the first pack.

Question: How many complete folders can Ms. Patel make with the 2 packs of paper? Show your work or explain your answer.

First, determine how much paper Ms. Patel has. The equation

$85 + 85 + 33 = ?$ can be used to solve.

Note that the second pack of paper does not contain 33 sheets of paper, but contains 33 more sheets than the first pack of 85. Ms. Patel has a total of 203 sheets of paper.

Next, determine how many complete folders can be made with 203 sheets of paper. Each folder needs to include 9 sheets of paper. The equation $203 \div 9 = ?$ can be used to solve.

By first seeing that $9 \times 20 = 180$, we know that she can so far make 20 folders, and there are 23 sheets of paper left.

Then, we see that $9 \times 2 = 18$, so she can make 2 more folders.

There will be 5 sheets of paper left, which cannot be used to make another complete folder. We cannot use the remainder, so we remain with 20 + 2 folders.

Answer: Ms. Patel can make 22 complete folders with the paper:

$$85 + 85 + 33 = 203 \text{ and } 203 \div 9 = 22 \text{ with a remainder of } 5$$

Solving Problems Using Letters to Represent Unknown Values

Early on you solved problems in which the unknown value looked like this:

$$2 + 3 = \_\_\_\_\_\_ \qquad 4 + 5 = ?$$

We can use a letter to represent what is unknown. The equation may look like this:

$$2 + 3 = X \qquad 4 + 5 = N$$

This simply means when we solve for X or N, we find the value that it represents.

This is the same thing as finding the answer.

If $2 + 3 = X$ and $2 + 3 = 5$, then $X = 5$.

We substitute the value of 5 for the letter X because we now know what X represents.

If $4 + 5 = N$ and $4 + 5 = 9$, then $N = 9$.

Sometimes the letter may appear before the equal sign. For example,

$Y + 7 = 15$. To solve this, ask yourself "What number, when added to 7, equals 15?"

We can subtract to solve. Since $15 - 7 = 8$, then $8 + 7 = 15$.

Therefore, $Y = 8$.

We can check by substituting 8 for Y. If we are correct, the result will be 15:

$$8 + 7 = 15, \text{ so } Y = 8$$

Here is another example:

$$5 \times N = 40$$

When we multiply, we use the letter X as the multiplication sign, so we avoid using the letter X as the unknown value.

Ask yourself, "What number, when multiplied by 5, equals 40?"

Since we know that $5 \times 8 = 40$, then $N = 8$.

We can also use division to solve the equation above.
If $5 \times N = 40$, then $40 \div 5 = N$.

Solving Multistep Problems Using Letters

Another way to solve multistep problems is to use a letter when finding an equation to help solve the problem. The letter is usually something that helps us understand what is unknown in the problem.

Example 1

Isabella did her math homework last night. She answered 12 questions. It took her 3 minutes to answer each question.

Question: What was the total number of minutes that it took Isabella to finish her math homework? Write an equation using a letter to stand for the number of minutes. Use the letter *m* to stand for the number of minutes:

$$12 \times 3 = m$$

Since $12 \times 3 = 36$, then $m = 36$.

Answer: $12 \times 3 = m$, $m = 36$. It took Isabella a total of 36 minutes to finish her math homework.

Assessing Reasonableness Using Mental Computation

When solving problems using the four operations, we can check to see if our answers make sense by rounding the numbers to estimate the answer. If the answer is close, or reasonable, our answer may be correct.

Example 1

Solve $63 \times 78 = ?$. To check, round and estimate the answer.

$60 \times 80 = 4{,}800$ is the estimated answer. The actual answer is 4,914.

$$\begin{array}{r} 78 \\ \times\ 63 \\ \hline 234 \\ +\ 4{,}680 \\ \hline 4{,}914 \end{array}$$

4,800 is close, so the answer of 4,914 is reasonable.

Example 2

Solve. Jonas had 485 marbles. Out of the 485 marbles, 98 are one color. The rest are multicolored. How many multicolored marbles does Jonas have?

Solve 485 – 98 = ?.

To check, round and estimate the answer:

$$500 - 100 = 400$$

The actual answer is 387. Since 400 is close, 387 is reasonable.

Example 3

Solve. There are 27 boxes of blueberries. Each box has 59 blueberries. How many total blueberries are in the 27 boxes?

Solve

$$\begin{array}{r} 59 \\ \times\ 27 \\ \hline 413 \\ +\ 118 \\ \hline 531 \end{array}$$

(There is a mistake here, this should be 1,180.)

To check this answer, round and estimate:

$$30 \times 60 = 1{,}800$$

The answer 531 is not reasonable, so there is a mistake in solving the problem.

Finding Factor Pairs and Factors of Numbers

Factors can be found in pairs or in a list.

Tip: Find the pairs in numerical order.

Example 1

Find the factor pairs of 20.

20

1 × 20 (1 is a factor of every number and is paired with the number.)

2 × 10 (All even numbers have a factor of 2.)

4 × 5 (Since the next number after 4 is already on the right side, there are no more.)

Example 2

Find the factors of 28.

First, find the pairs:

$$28$$
$$1 \times 28$$
$$2 \times 14$$
$$4 \times 7$$

Next, place the numbers in a list in order:

1, 2, 4, 7, 14, and 28 are the factors of 28.

Prime and Composite Numbers

To determine whether a number is prime or composite, find the factors.

Factors of 7

Example 1

Only $1 \times 7 = 7$

7 is prime because it has only one factor pair (1 and itself).

Factors of 9

Example 2

1×9 and $3 \times 3 = 9$

9 is composite because it has more than one factor pair.

Did You Know?
The number 1 is not prime and not composite.

Prime and Composite Numbers Up to 20

2, 3, 5, 7, 11, 13, 17, and 19 are prime.

4, 6, 8, 9, 10, 12, 14, 15, 16, 18, and 20 are composite.

Patterns

A pattern is a set of numbers or objects that are arranged and related based on a specific rule.

Number Patterns

When the rule is given, you can extend the pattern.

When the rule is not given, it must be determined before the pattern can be extended.

Example 1

Extend the pattern using the rule "add 3":

1, 4, 7, ____, ____, ____

7 + 3 = 10; 10 + 3 = 13; 13 + 3 = 16

1, 4, 7, 10, 13, 16

Example 2

Find the rule and extend the pattern:

5, 9, 13, 17, ____, ____, ____

To find the rule, first find the rule's operation.

In this pattern the rule is to add.

Next, find the difference between two numbers next to each other in the pattern:

9 – 5 = 4   13 – 9 = 4   17 – 13 = 4

The rule is "add 4" because the difference between the numbers is 4 and the numbers are increasing.

Finally, apply the rule to extend the pattern:

17 + 4 = 21; 21 + 4 = 25; 25 + 4 = 29

5, 9, 13, 17, 21, 25, 29

Example 3

Find the rule and extend the pattern:

72, 63, 54, 45, ____, ____, ____

The numbers are decreasing and the operation is subtraction.

The difference between the numbers is 9:

$$72 - 63 = 9;\ 63 - 54 = 9;\ 54 - 45 = 9$$

Apply the rule "subtract 9" to extend the pattern:

72, 63, 54, 45, 36, 27, 18

Shape Patterns

When extending shape patterns, look for the change in the shapes. It could be in the number of sides, or the shading.

The example below shows a change in the shading:

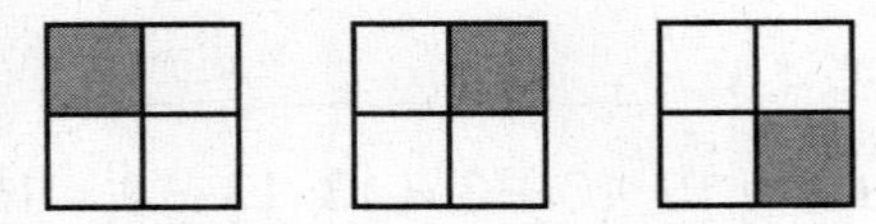

The shaded area goes from top left, top right, bottom right. This follows a clockwise movement. The next shape should be shaded in the bottom left:

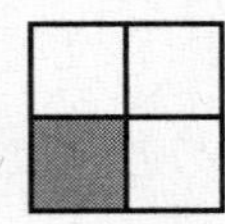

Identify Additional Features of Number Patterns

With some patterns, other traits can be determined such as whether the numbers are odd, even, prime, or composite.

Example 1

The patterns below follow the rule "add 2":

Pattern 1: 7, 9, 11, 13, 15, 17

Pattern 2: 4, 6, 8, 10, 12

Pattern 1 has only odd numbers. Pattern 2 has only even numbers.

Since every other number is even and odd, patterns following the rule "add 2" will have the same type of number (odd/even) that starts the pattern.

Example 2

The patterns below follow the rule "add 3":

Pattern 3: 5, 8, 11, 14, 17

Pattern 4: 6, 9, 12, 15, 18

Both patterns alternate between odd and even numbers no matter what the pattern starts with.

Example 3

The pattern below follows the rule "subtract 5":

23, 18, 13, 8, 3

Question: How many prime numbers are in the pattern?

Answer: The numbers 23, 13, and 3 are prime numbers. There are three prime numbers in the pattern.

Operations and Algebra Practice Problems

1. Justin started a baseball card collection last year. At the end of last year, he had a total of 8 cards. This year, he has 7 times as many cards as last year's total.

   Place the numbers and symbols shown below into the boxes to correctly complete an equation that can be used to find the total number of cards in Justin's collection this year.

1	7	8	15	49	56	64	+	−	×	÷

   ☐ ☐ ☐ = ☐

2. Tyler and Jarrad's dad is 36 years old. Their dad is 3 times as old as Tyler and 4 times as old as Jarrad.

Circle the numbers in the statements shown below to make them true:

Dad is 36 years old. Tyler is _____ years old. Jarrad is _____ years old.

Tyler	Jarrad
3	4
8	8
9	9
12	12

3. Look at the array shown below:

x x x x
x x x x

Can this array be used to solve the equation $4 \times 4 = ?$

Circle one: Yes or No

Why or why not? --

----

4. The diagram below can be used to show the factors of 24 and 30. Use the numbers below to correctly complete the diagram. Use each number once.

3	4	5	6	8	10	12	15

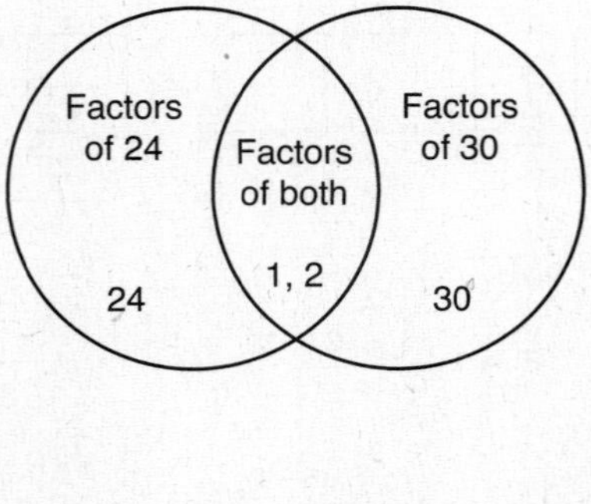

Factors of 24

24

Factors of both

1, 2

Factors of 30

30

5. There are 20 girls and 25 boys marching in the band. They are marching in 5 rows. Which of the following shows the number of band members marching in each row?

○ A. 4
○ B. 5
○ C. 9
○ D. 10

6. Which four numbers are prime numbers?

☐ A. 2
☐ B. 7
☐ C. 9
☐ D. 11
☐ E. 16
☐ F. 19

7. Alejandro's mom brought 18 juice packs to the class picnic. Each juice pack contains 8 juice boxes. The students drank a total of 98 juice boxes at the picnic.

PART A

Write an equation to find the total number of juice boxes left after the picnic. Let *b* represent the number of juice boxes left. Solve your equation.

PART B

Find the number of full juice packs that can be made from the remaining juice boxes. Show your work or explain your answer.

8. Answer the following statements by circling T (true) or F (false).

A. 16 is a multiple of 2 T F
B. 20 is a multiple of 3 T F
C. 24 is a multiple of 9 T F
D. 28 is a multiple of 4 T F
E. 32 is a multiple of 6 T F
F. 36 is a multiple of 8 T F
G. 42 is a multiple of 7 T F
H. 55 is a multiple of 5 T F

9. Francesca started a number pattern with the rule "add 7." Use the numbers shown below to complete Francesca's pattern. Some numbers will not be used.

7	10	14	17	21	24	31

3, _____, _____, _____, _____

10. Nicolas wants to place square foot tiles on the floor in his room.

- He has a total of 32 square tiles.
- He wants to use all of the tiles.
- He wants to place them in 4 equal-sized rows.

Which of the following could be the number of tiles in each of the 4 rows?

O A. 8 tiles
O B. 9 tiles
O C. 16 tiles
O D. 28 tiles

11. Katya wants to construct a number pattern that starts with the number 2. She wants the pattern to follow the rule "add" but has not decided what number to add for her rule.

2, ____, ____, ____, ____

Circle the response T (true) or F (false) to correctly complete each statement.

A. Adding an even number will result in a pattern with only even numbers.

T F

B. Adding an odd number will result in a pattern with only odd numbers.

T F

C. Adding an even number will result in a pattern with alternating even and odd numbers.

T F

D. Adding an odd number will result in a pattern with alternating even and odd numbers.

T F

12. The fourth-grade students are all receiving 2 folders: one for their homework and one for their tests. The chart shows the classes and the number of students in each class.

Class	Number of Students
4A	22
4B	24
4C	21

PART A

Write an equation to show the total number of fourth-grade students using the letter *s* to stand for the number of students. Solve your equation.

PART B

The folders come in packs of 8. How many packs will they need so that every student receives 2 folders? Show your work or explain your answer.

13. Rosie and her family went to Sunset Lake for the weekend. On Saturday, there were a total of 456 people at the lake. On Sunday, there were 38 fewer visitors than on Saturday.

Which of the following shows the total number of people at the lake on Saturday and Sunday?

- O A. 418
- O B. 494
- O C. 874
- O D. 950

14. Lenny had 14 boxes of baseballs. There were 8 baseballs in each box. At the game, 13 baseballs were hit over the foul ball fence and were lost.

Use the numbers below to complete the equation and the statement to find how many baseballs Lenny has left after the baseball game:

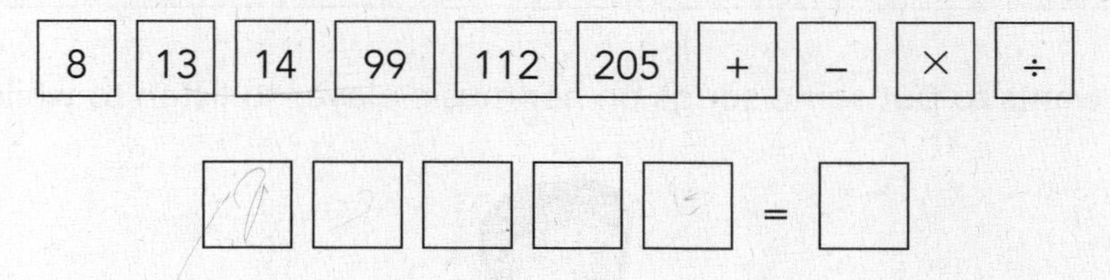

There were _____ baseballs left after the game.

15. Ava completed some multiplication problems by estimating the answers. The chart below shows the problems and Ava's estimated answers.

Circle Y (yes) or N (no) to show if Ava correctly estimated her answers to the problems.

Multiplication Problem	Estimated Answer	Reasonable Estimate? Y (Yes) or N (No)
A. $79 \times 32 = ?$	2,400	Y N
B. $19 \times 410 = ?$	8,000	Y N
C. $498 \times 6 = ?$	30,000	Y N
D. $915 \times 4 = ?$	3,600	Y N
E. $205 \times 100 = ?$	2,000	Y N
F. $2{,}995 \times 3 = ?$	9,000	Y N

16. Sofia wants to put some songs on her music player to listen to while she jogs.

- She wants the total amount of music to be no less than 40 minutes and no more than 60 minutes.
- She wants to choose songs that are no less than 6 minutes and no more than 8 minutes each.

What could be a possible number of songs and length of each song that Sofia can put on her music player? Show your work or explain your answer.

Answers to Operations and Algebra Practice Problems

1. $8 \times 7 = 56$

2. Tyler is 12 years old: $12 \times 3 = 36$
   Jarrad is 9 years old: $9 \times 4 = 36$

3. No. The array shows 2 rows of 4 to equal 8. For 4×4, you need 4 rows of 4 to equal 16.

4. Factors of 24: 4, 8, 12
   Factors of 30: 5, 10, 15
   Factors of both: 3, 6

5. ***C*** $20 + 25 = 45$
   $45 \div 5 = 9$

6. **A** 2 **B** 7 **D** 11 **F** 19

7. Part A: $18 \times 8 - 98 = b$

   $144 - 98 = 46$, $b = 46$

   Part B: $46 \div 8 = 5$ R6
   5 full packs

8. A. T
   B. F
   C. F
   D. T
   E. F
   F. F
   G. T
   H. T

9. Pattern: 3, 10, 17, 24, 31

10. **A** 8 tiles
    $32 \div 4 = 8$

11. A. T: A pattern of + 2 would be 2, 4, 6, 8, 10.
    B. F: A pattern of + 3 would be 2, 5, 8, 11, 14.
    C. F
    D. T

12. Part A: $22 + 24 + 21 = s$.
    The solution is $s = 67$.
    Part B: 17 packs of folders
    A total of 67 students × 2 folders each = 134
    134 ÷ 8 = 16 R6.
    17 packs are needed.

13. ***C*** 874
    Saturday: 456 people at Sunset Lake
    Sunday: 456 – 38 = 418 at Sunset Lake
    Total visitors on Saturday and Sunday: 456 + 418 = 874

14. $14 \times 8 - 13 = 99$.
    There are 99 baseballs left after the baseball game.

15. A. Y
    B. Y
    C. N
    D. Y
    E. N
    F. Y

16. Possible number of songs and lengths:

    5 songs, 8 minutes each (40 minutes)
    6 songs, 7 minutes each (42 minutes)
    6 songs, 8 minutes each (48 minutes)
    7 songs, 6 minutes each (42 minutes)
    7 songs, 7 minutes each (49 minutes)
    7 songs, 8 minutes each (56 minutes)
    8 songs, 6 minutes each (48 minutes)
    8 songs, 7 minutes each (56 minutes)
    9 songs, 6 minutes each (54 minutes)
    10 songs, 6 minutes each (60 minutes)

Chapter Checklist for Standards

If you answered the questions correctly, you are on your way toward mastering the concepts and skills for this standard.

Standard	Concept/Skill	Questions
4.OA.1	I understand that multiplication is comparing groups or amounts in groups.	1, 2
4.OA.2	I can solve word problems by writing equations and using multiplication to find a missing or unknown value.	3, 5, 7, 10, 16
4.OA.3	I can use any and all of the four operations to solve multistep word problems with whole numbers. I can write equations and use letters to represent the unknown quantity. I can round numbers to estimate in order to determine if my answer is reasonable.	12, 13, 14, 15
4.OA.4	I can find all the factors of whole numbers less than 100. I can determine if a number is prime or composite.	4, 6, 8
4.OA.5	I can find the rule for a number or shape pattern in order to extend it. I can also create a pattern.	9, 11

Numbers and Operations in Base Ten (NBT)

Our number system is a base ten system. This means that we use ten different digits to form our numbers. Ever since kindergarten, we have used numbers for counting and measuring. Let's review some key terms.

VOCABULARY

Digit: single symbols used for writing numbers; 0, 1, 2, 3, 4, 5, 6, 7, 8, 9 are digits

Numbers: a count, an amount, or a measurement made up of digits; the number 358 uses the digits 3, 5, and 8.

Place value: the positions of the digits that determine its numerical worth or value

By fourth grade, you should be familiar with place value up to the millions place:

____ ,	____	____	____ ,	____	____	____
millions	hundred thousands	ten thousands	thousands	hundreds	tens	ones

Value: a numerical worth of a digit based on its position in a number; the 3 in 358 has a value of 300 because it is in the hundreds place, the 5 in 358 has a value of 50 because it is in the tens place, and the 8 in 358 has a value of 8 because it is in the ones place

What Does This Mean?
We write numbers using digits. Where we place the digits determines the value of the digits and of the number.

Reading and Writing Numbers

We place digits in groups of one and three and place commas between the groups. Only the group on the left of the comma can have less than three digits. Once you start a number, all of the groups to the right have to have three digits between each comma. We pronounce the number in separate groups.

Example

1,234,567 is pronounced "one million, two hundred thirty-four thousand, five hundred sixty-seven" (we don't pronounce the "ones" group).

Pronounce each group as a 3-digit number, then say the group (except the ones).

Placing Commas

We need to place commas with numbers that have more than 3 digits. We count every 3 digits from right to left and place a comma:

___ , ___ ___ ___ , ___ ___ ___

COMMON ERROR

Avoid counting from the left when placing commas:

Incorrect: 234,23 Correct: 23,423

> **Note**
> If we move the same digit one place to the left, the value of the digit increases 10 times. In the same manner, moving to the right decreases the value 10 times.

Example

Compare the digit 4 in the numbers **4**,270 and **4**2,700.

In the number 4,270 the value of the 4 is 4,000.

In the number 42,700 the value of the 4 is 40,000.

40,000 is 10 times more than 4,000 (4,000 × 10 = 40,000).

Number Forms

There are three forms that we use to read and write numbers.

1. **Standard form:** This is the most common form for numbers.
2. **Word form:** This is the number written as words and is the same as the number is pronounced when spoken.
3. **Expanded form:** This is the number "broken apart" to show the value of each digit in the number.

Example 1

Standard Form	Word Form	Expanded Form
1,358,729	One million, three hundred fifty-eight thousand, seven hundred twenty-nine	1,000,000 + 300,000 + 50,000 + 8,000 + 700 + 20 + 9

When you check the standard and word form of the same number, read the word form while looking at the standard form to see if they are the same number.

When you check the expanded form of a number, the parts are the separate values, and they should add up to the standard form of the same number.

Example 2

Standard Form	Word Form	Expanded Form
406,850	Four hundred six thousand, eight hundred fifty	400,000 + 6,000 + 800 + 50

Notice that in this example, the digit 0 is in the ten thousands and in the ones place in the standard form, so there is no value for these digits in the expanded form.

Comparing Numbers

When comparing numbers, we compare the values of a pair of numbers.

One number is either greater than (>), less than (<), or equal to (=) another number. We place the sign (>, <, or =) between the two numbers. To compare numbers, we ask ourselves:

1. **Do both numbers have a different number of digits?** If so, the number with the greater number of digits is larger.

   98,244 ? 115,013

   Even though the digit 9 may be larger than the digit 1, the value is not larger because the 1 is in the hundred thousands place and the 9 is in the ten thousands place:

   98,244 < 115,013 because 90,000 < 100,000

2. **Do both numbers have the same number of digits?** If so, then compare the value of the greatest digit:

   35,078 ? 28,705

   The value of the ten thousands in the first number is 30,000 and the value of the ten thousands in the second number is 20,000.

   So, 35,078 > 28,705.

3. **Do both numbers have the same number of digits and start with the same digit?** If so, keep looking to the right until you see different digits in the same place, and then compare their values:

   519,042 ? 518,420

   The same digit is in both the hundred thousands place (5) and the ten thousands place (1), so we keep looking and see that in the thousands place 9,000 > 8,000.

   So, 519,042 > 518,420.

Note

790,663 = 790,663 because both numbers have the same number of digits and have the same digits in the same places. They have the same value.

Ordering Numbers

When there are three or more numbers, we can order the numbers by placing them in order either from the least to the greatest value or from the greatest to the least value.

Example

Place the numbers below in order from least to greatest:

506,779   490,813   90,887   98,001

1. Look at the directions and check for the order (least to greatest).

If there are different numbers of digits, start with the least number of digits.

2. Check for the values and place those with the same number of digits in order first:

90,887   98,001   (90,000 is less than 98,000)

3. Look for the numbers with the larger number of digits and place them in order following the same directions:

490,813   506,779

4. Finally, place the entire number set in order according to the directions:

90,887   98,001   490,813   506,779

COMMON ERROR

Not reading the directions and assuming the order without reading.

Example

506,779   490,813   98,001   90,887

This set of numbers is in order from greatest to least. If the directions stated "least to greatest," this response would be incorrect.

Rounding Numbers and Estimating

We round numbers by finding a number that is close to the exact amount.

We can use a number line to find the closest number.

Round 32,628 to the nearest ten thousand.

32,628 is between 30,000 and 40,000 when counting by ten thousands.

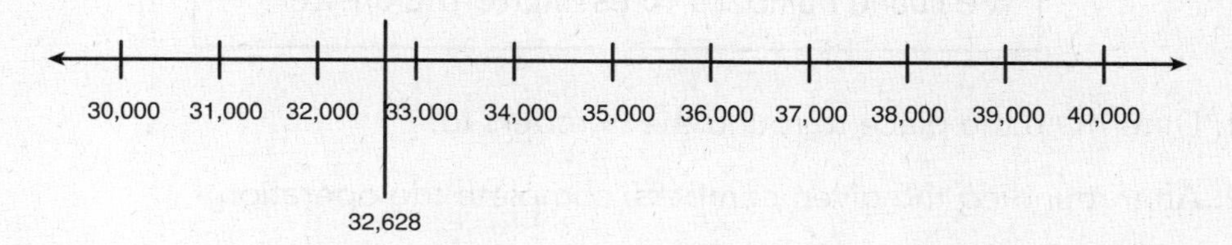

We can see that 32,628 is closer to 30,000, so we round back or stay at 30,000.

Round 32,628 to the nearest thousand. 32,628 is between 32,000 and 33,000 when counting by thousands.

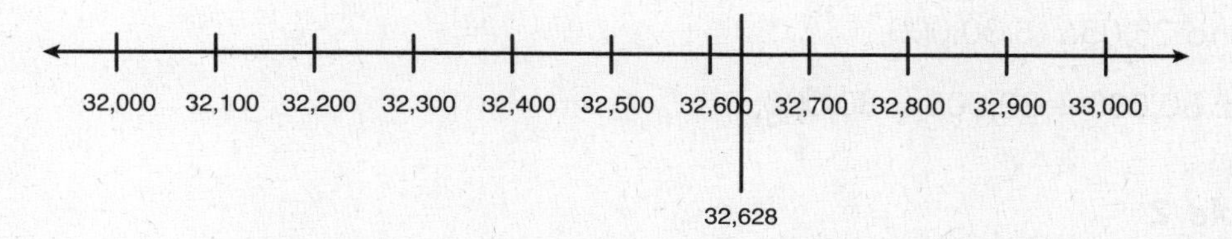

We can see that 32,628 is closer to 33,000, so we round up to 33,000.

Strategy for Rounding Numbers

1. Decide which place to round the number.
2. Find the closest rounded numbers before and after your exact number when counting by that place.
3. Check the digit to the right of the place you are rounding to. Round up to the next number if the digit to the right is 5, 6, 7, 8, or 9. Round back to the number if the digit to the right is 0, 1, 2, 3, or 4.

Note

For single-digit numbers, do not round them, just leave them alone. We do not want to round any number to a value of 0.

Rounding Numbers and Estimating Sums and Differences

We round numbers so that we can perform operations using mental math instead of using pencil and paper or a calculator. When we round numbers and then calculate the answer, we are estimating: finding a value that is close but not the exact answer.

Note

We round numbers to estimate the answer.

Step 1: Determine the place to round the numbers to.

Step 2: After rounding the given numbers, complete the operation.

Example 1

Find the estimated sum of 52,433 and 28,056 by rounding to the nearest ten thousand.

Round 52,433 to 50,000.

Round 28,056 to 30,000.

Add: 50,000 + 30,000 = 80,000.

Example 2

Find the estimated difference of 127,998 and 81,575 by rounding to the nearest ten thousand.

Round 127,998 to 130,000.

Round 81,575 to 80,000.

Subtract: 130,000 – 80,000 = 50,000.

COMMON ERROR

If the directions state "round before solving" or "estimate the (sum/difference/product/quotient)," do not find the actual answer and round the answer. You should ROUND first, then SOLVE.

Here is an example:

Round the numbers to the nearest thousand to find the estimated sum:

Incorrect	Correct
2,463	2,463 rounds to 2,000
+ 3,378	3,378 rounds to + 3,000
5,841 , which rounds to 6,000	5,000

The estimated sum can be confused as both 6,000 and 5,000.

When the directions are to round first, then solve, the estimated sum is 5,000.

Rounding Numbers to Find Estimated Products and Quotients

We use the same rounding rules for finding products.

Example

Round to the nearest ten (58) and nearest hundred (712) to find the estimated product of 58 and 712.

Round 58 to 60.

Round 712 to 700.

Solve 60×700.

Strategy for Multiplication

1. Find and solve the fact: $6 \times 7 = 42$.
2. Count and add on the additional zeros. Since there are 3 more zeros, our product is 42,000.

COMMON ERROR

If the answer to the fact ends in a zero, do not count that zero when adding on the additional zero.

53×625 rounds to 50×600.

$5 \times 6 = 30$, then add on the 3 extra zeros.

$50 \times 600 = 30{,}000$. (Even though it looks like 4 zeros, the zero in the 30 is part of the fact.)

The estimated product is 30,000 not 3,000.

Estimating Quotients

When estimating quotients, we do not follow regular rounding strategies.

We round numbers so that we can use mental math to divide; therefore, we need to round numbers so that they can be easily divided.

Find the estimated quotient of 312 and 8.

If we used regular rounding rules, we would not be able to easily solve $300 \div 8$. We want to find the closest number to 312 that can be easily divided by 8.

312 can be rounded to 320, which is easily divided by 8:

$$8 \times ? = 320$$

$$320 \div 8 = 40$$

Addition of Whole Numbers

When we add, we combine two or more values to find the sum or total of the values.

We change the value of one number by finding the difference between that number and another number.

The values we add are called "addends":

addend + addend = sum

We can use the number line as a visual model of addition.

Example 1

$33 + 8 = ?$

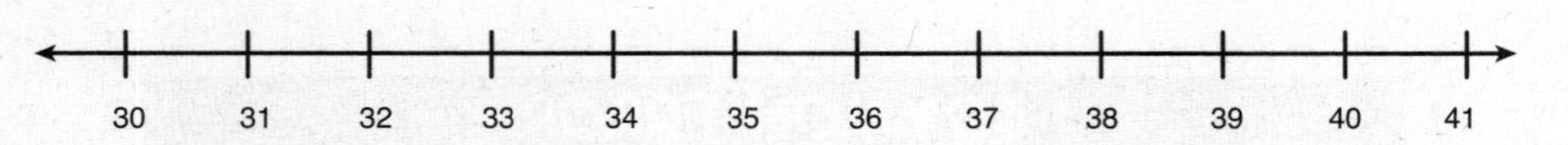

When we add 33 + 8, we start from the value of 33 and count up 8 more to find the total value of 41.

Example 2

What number is 12 more than 75?

We add 75 + 12 to find the number that is 12 more than 75:

$$75 + 12 = 87$$

> **Note**
> The commutative, or order property, allows us to change the order of the two addends and still have the same sum.

When we add 8 + 33, we also find the sum of 41:

$$33 + 8 = 8 + 33$$

The associative, or grouping property of addition, allows us to add 3 or more values by using parentheses to first add two numbers, then add the third to find the total sum. When we change the order, the sum remains the same:

$$(17 + 5) + 8 = 22 + 8 = 30$$

$$17 + (5 + 8) = 17 + 13 = 30$$

$$(17 + 5) + 8 = 17 + (5 + 8)$$

When we add multidigit numbers we stack the numbers vertically, making sure to line up the number places. When the sum of two digits is greater than 10, we carry the number in the tens place to the next place.

Example 1

$$\begin{array}{r} 5,7{}^{1}05 \\ +\,3,019 \\ \hline 8,724 \end{array}$$

In the ones place, 5 + 9 = 14, we place the 4 in the ones place and carry the 1 to the tens place

Example 2

Find the sum of 6,053 and 897.
Write the numbers vertically, lining up the place value columns correctly:

$$\begin{array}{r} 6,{}^{1}0{}^{1}53 \\ +\quad 897 \\ \hline 6,950 \end{array}$$

COMMON ERROR

If the columns are not lined up correctly, the result is an incorrect sum.

$$\begin{array}{r} 6{,}053 \\ +\ 8\ 97 \\ \hline 15{,}023 \end{array}$$

In this problem, the 897 is incorrectly placed. The value of the number is really 897, but the values are added as 8,970.

Subtraction of Whole Numbers

When we subtract:

- We look for the difference between two values.
- We decrease a number by the value of the second number.
- We break apart a total value and find the unknown part.

We can use the number line as a visual model for subtraction:

61 – 13 = ?

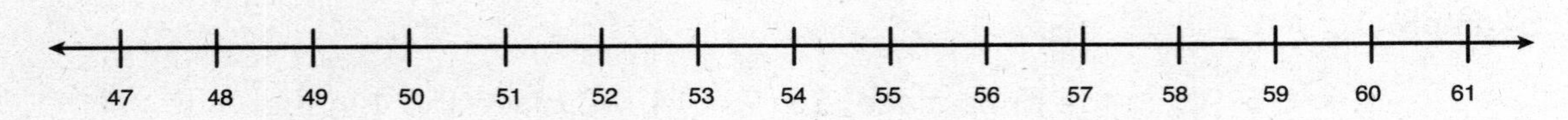

When we subtract 61 – 13, we start with the total value of 61 and take away by counting back 13 spaces on the number line to 48. The value 61 is 13 more than 48:

61 – 13 = 48

We can check our answer by adding 13 + 48 = 61.

This is due to the relationship between addition and subtraction. We can find the missing part of the subtraction problem by using addition.

When we subtract multidigit numbers we stack the number vertically, making sure to line up the number places.

$$\begin{array}{r} 9{,}3\ {}^{6}\cancel{7}\ {}^{11}\cancel{1} \\ -\ 7{,}2\ \ 6\ \ \ 4 \\ \hline 2{,}1\ \ 0\ \ \ 7 \end{array}$$

In the ones place, we cannot solve 1 – 4. We rename 7 tens to 6 tens and the 1 becomes 11 ones, so we can subtract 4 from 11 ones. We only have to rename the tens and ones here because there are enough hundreds and thousands.

WHY DOES THIS WORK?

In this subtraction problem, we start with a value of 9,371.

When we rename the number, the value is still 9,371, but the digits are different. We borrow from the next place to rename the values in each place:

$$9{,}371 = 9{,}000 + 300 + 70 + 1$$

9,371 is renamed as 9,000 + 300 + 60 + 11, but it still equals 9,371.

$$\begin{array}{rrrrl} 9{,}000 & 300 & 60 & 11 & \\ -\ 7{,}000 & 200 & 60 & 4 & \\ \hline 2{,}000 & 100 & 0 & 7 & = 2{,}107 \end{array}$$

COMMON ERROR

We cannot subtract from the bottom up.

$$\begin{array}{r} 9{,}3\ 7\ 1 \\ -\ 7{,}2\ 6\ 4 \\ \hline 2{,}1\ 1\ 3 \end{array}$$

We cannot subtract 4 from 1 in the ones place. We must borrow from the tens place to subtract 11 – 4.

Multiplication of Whole Numbers

When we multiply:

- We find the total of equal groups by repeatedly adding the group totals.

5×3

= 15

- We resize one number by the value of another number.
- We can use the number line as a visual model for multiplication:

$3 \times 5 = ?$

0 1 2 3 4 5 6 7 8 9 10 11 12 13 14 15 16

- We can skip count by 3s.
- We can add 3 + 3 + 3 + 3 + 3.
- We see that the number 15 is 5 times as many as 3:

$$3 \times 5 = 15$$

- We can find the area of a rectangle that is 5 units by 3 units:

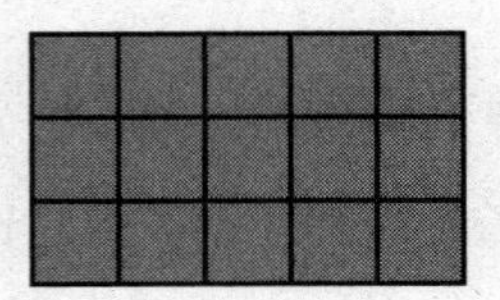

We multiply multidigit numbers by stacking the numbers vertically and multiply based on place value.

$$\begin{array}{r} 328 \\ \times \quad 5 \\ \hline 1{,}640 \end{array}$$

When we break this apart, we can see that we are solving

$(5 \times 8) + (5 \times 20) + (5 \times 300)$

$40 + 100 + 1{,}500 = 1{,}640$

$$\begin{array}{r} 62 \\ \times \ 37 \\ \hline 434 \\ +\ 1{,}860 \\ \hline 2{,}294 \end{array}$$

We begin by solving $(7 \times 2) + (7 \times 60)$

$14 + 420 = 434$

Then we solve $(30 \times 2) + (30 \times 60)$

$60 + 1{,}800 = 1{,}860$

Finally, we solve $434 + 1{,}860 = 2{,}294$.

COMMON ERROR

In the above problem, when you multiply 3 × 62 you are really multiplying 30 × 62, so there must be a 0 in the ones place.

$$\begin{array}{r} 62 \\ \times \ 37 \\ \hline 434 \\ +\ 186 \\ \hline 620 \end{array}$$

← **This is incorrect, it should be 1,860.**

Division of Whole Numbers

When we divide:

- We start with the total amount (dividend) and find the number of equal-sized groups (divisor), or the number in each equal-sized group.
- We find the most amount of equal-sized groups there could be in a total, or the most amount that can be placed in equal-sized groups (quotient):

$$24 \div 3 = 8$$

- We can use the number line as a visual model for division.

Example 1

$24 \div 8 = ?$

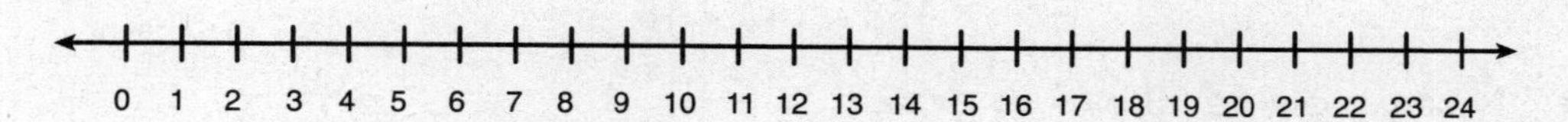

- We can start with 24 and use repeated subtraction to see how many groups of 8 are in 24. We can subtract 8 a total of 3 times to go from 24 to 0:

$$24 \div 8 = 3$$

Example 2

$22 \div 6 = ?$

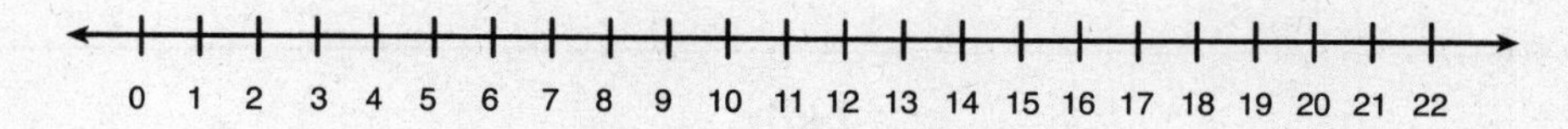

We see a total of 3 groups, but this leaves us with a remainder of 4.

$22 \div 6 = 3$ with a remainder of 4.

When we divide multidigit numbers, we set up the problem vertically and divide based on place value until we can no longer divide:

$$\begin{array}{r} 122 \text{ R4} \\ 7\overline{)858} \\ \underline{7} \\ 158 \\ \underline{14} \\ 18 \\ \underline{14} \\ 4 \end{array}$$

- Begin by dividing 800 by 7 to get 100:

$$100 \times 7 = 700$$
$$858 - 700 = 158$$

- Next we divide 150 by 7 to get 20:

$$20 \times 7 = 140$$
$$158 - 140 = 18$$

- Next we divide 18 by 7 to get 2:

$$2 \times 7 = 14$$
$18 - 14 = 4$, which becomes the remainder.

We can check our division problem with multiplication and addition.

If $858 \div 7 = 122$ with a remainder of 4, that means we can make 122 equal groups with 4 left.

$$7 \times 122 = 854 + 4 = 858$$

Note

Long division takes a lot of steps combining division, multiplication, and subtraction. Be sure to check all work to avoid mistakes.

COMMON ERRORS

- Mistakes in multiplication facts
- Subtraction errors
- Remainders larger than the dividing by number

Numbers and Base Ten Practice Problems

1. The word form of a number is shown below:

Nine hundred seventy-five thousand, four hundred six

PART A

Show the standard form of the number by placing the digits in the boxes in the correct place. All of the digits may not be used.

0	1	2	3	4	5	6	7	8	9

_____ _____ _____ , _____ _____ _____

PART B

Write the expanded form of the number:

---

2. The standard form of a number is shown below:

12,043

PART A

Circle the correct value to show the expanded form of the number.

_____ + _____ + _____ + _____

1,000	20,000	400	3
10,000	2,000	40	30
100,000	200	4	300

PART B

Which of the following is the correct word form of the number 12,043?

O A. Twelve hundred, forty-three
O B. Twelve thousand, forty-three
O C. Twelve thousand, four hundred three
O D. Twelve hundred thousand, four hundred three

3. Complete the chart below by placing numbers in the correct boxes based on the value of the digit 3 in each of the numbers. Some of the numbers may be placed in more than one box.

312,055 213,505 313,929 430,672

132,550 133,299 309,412

Numbers with the Digit 3 in the Hundred Thousands Place	Numbers with the Digit 3 in the Ten Thousands Place	Numbers with the Digit 3 in the Thousands Place

4. The standard form of a number is shown below:

79,653

Write a number in which the value of the digits 9 and 5 are both 10 times more than the value of the digits 9 and 5 in the number shown above:

----

5. The expanded form of a number is shown below:

1,000,000 + 40,000 + 700 + 8

Which of the following is the correct standard form of the number?

O A. 1,040,708
O B. 1,040,780
O C. 1,400,708
O D. 1,400,780

6. Ella and Maria each wrote a number as shown:

Ella's Number	Maria's Number
418,502	145,820

Which three statements are true about the digits in Ella's and Maria's numbers? Choose THREE that are correct.

☐ A. The value of the digit 4 in Ella's number is 10 times more than the 4 in Maria's number.

☐ B. The value of the digit 1 in Ella's number is 10 times more than the 1 in Maria's number.

☐ C. The value of the digit 5 in Maria's number is 10 times more than the 5 in Ella's number.

☐ D. The value of the digit 2 in Maria's number is 10 times more than the 2 in Ella's number.

☐ E. The value of the digit 8 in Maria's number is 10 times more than the 8 in Ella's number.

7. Different forms of some number pairs are shown in the chart below. Compare the values by circling the correct symbol. (<, =, or >) for each number pair.

Value of First Number	< = >	Value of Second Number
257,109	< = >	257,091
468,035	< = >	Four hundred sixty-eight thousand, three hundred five
609,212	< = >	600,000 + 90,000 + 200 + 10 + 2
80,000 + 5,000 + 300 + 60	< = >	Eighty-five thousand, three hundred sixty

8. The chart below shows some of New Jersey's largest cities and the population (number of people who live there).

City	Population
Elizabeth	127,558
Jersey City	257,342
Newark	278,427
Paterson	145,948
Toms River	88,791
Trenton	84,349

Place the names of the cities in the correct order to show the populations from least to greatest.

________ ________ ________ ________ ________ ________

least greatest

9. Complete the number sentences shown by placing the correct digit in the box. You will use each number one time.

2	3	7	9

A. 19,304 > 19,□98

B. 25,189 < 25,1□0

C. 32,400 > 32,□75

D. 57,626 = 5□,626

10. The chart shows some numbers and the rounded numbers when rounding to the nearest thousand. Circle Y if the rounded number is correct and N if the rounded number is not correct.

Number	Number Rounded to the Nearest Thousand	Rounded Correctly Y (Yes) or N (No)
A. 2,458	3,000	Y N
B. 16,349	16,000	Y N
C. 74,512	74,000	Y N
D. 99,199	100,000	Y N
E. 105,821	106,000	Y N
F. 232,077	232,000	Y N

11. A mystery number, when rounded to the nearest thousand, is 36,000. James says that the largest number possible that could be the mystery number is 36,400. Is he correct? Why or why not?

---

12. Solve:

A. $\begin{array}{r} 6{,}752 \\ +\ 2{,}548 \\ \hline \end{array}$

B. $\begin{array}{r} 9{,}044 \\ -\ 1{,}839 \\ \hline \end{array}$

C. $\begin{array}{r} 85 \\ \times\ 73 \\ \hline \end{array}$

D. $6\overline{)8{,}154}$

13. Complete the addition problem by placing the correct digit in the boxes. Some digits will not be used.

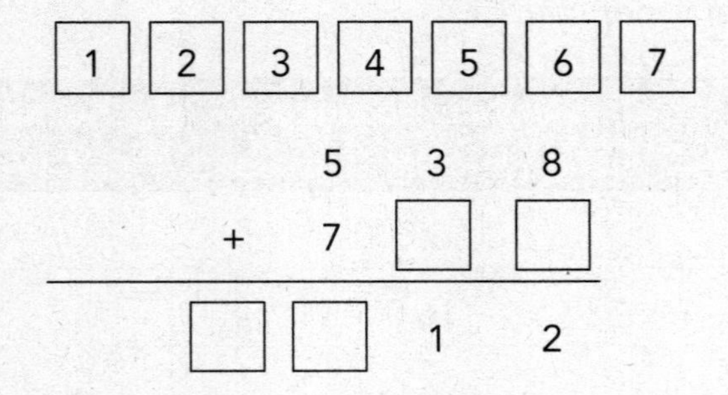

14. Nicole solved the subtraction problem as shown:

$$\begin{array}{r} 726 \\ -\ 549 \\ \hline 287 \end{array}$$

Did Nicole correctly solve the subtraction problem? Why or why not?

---

15. Olivia solved the following multiplication problem:

$$6 \times 3{,}142 = ?$$

Complete the number sentence below by placing the correct numbers in the boxes to show how Olivia could have solved the multiplication problem. Some numbers may be used more than once.

2	3	6	100	400	600	3,000	6,000	18,000

$$(6 \times \square) + (6 \times \square) + (\square \times 40) + (\square \times 2) = ?$$

16. Some division problems were solved. The problems and their quotients are shown:

A. $\begin{array}{r}940\\9\overline{)1{,}836}\end{array}$ C. $\begin{array}{r}2{,}041\\3\overline{)6{,}123}\end{array}$ E. $\begin{array}{r}408\\6\overline{)2{,}448}\end{array}$

B. $\begin{array}{r}407\\7\overline{)2{,}849}\end{array}$ D. $\begin{array}{r}1{,}180\\8\overline{)8{,}864}\end{array}$ F. $\begin{array}{r}503\\2\overline{)1{,}060}\end{array}$

Which THREE quotients are correct?

- ☐ A. Quotient A 940
- ☐ B. Quotient B 407
- ☐ C. Quotient C 2,041
- ☐ D. Quotient D 1,180
- ☐ E. Quotient E 408
- ☐ F. Quotient F 503

17. Place the digits in the boxes to correctly solve the equation shown. Some digits may be used more than once or not at all.

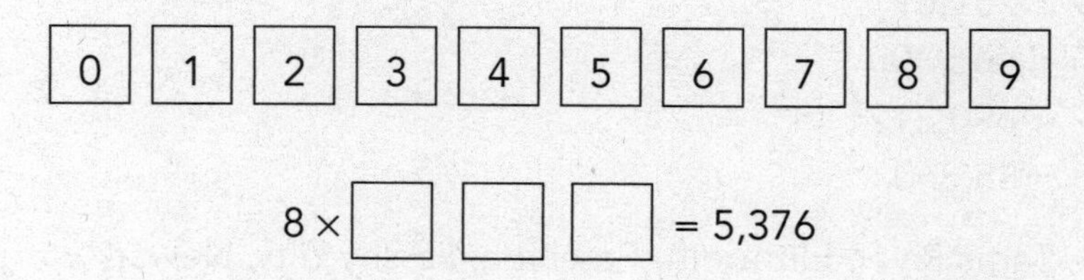

18. Place the digits in the boxes to correctly solve the equation shown. Some digits may be used more than once or not at all.

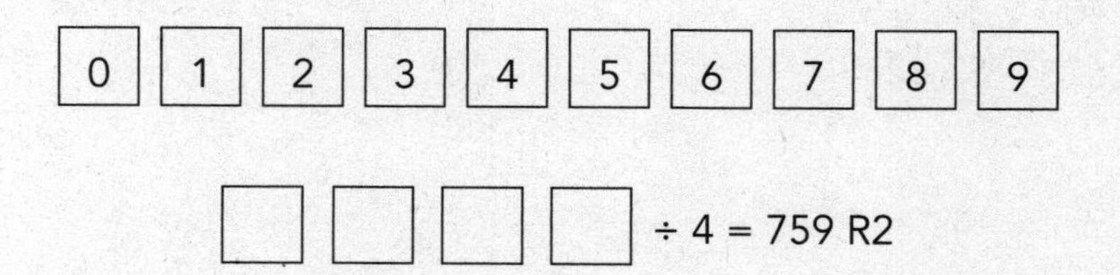

Answers to Numbers and Base Ten Practice Problems

1. Part A: 975,406
   Part B: 900,000 + 70,000 + 5,000 + 400 + 6

2. Part A: 10,000 + 2,000 + 40 + 3
   Part B: **B** twelve thousand, forty-three

3. Hundred thousands place: 312,055 313,929 309,412
   Ten thousands place: 132,550 133,299 430,672
   Thousands: 213,505 133,299 313,929

4. 97,563. Any number in which the 9 is in the ten thousands place and the 5 is in the hundreds place

5. **A** 1,040,708

6. **A** 40,000 × 10 = 400,000
   C 500 × 10 = 5,000
   D 2 × 10 = 20

7. 257,109 > 257,091
   468,035 < 468,305
   609,212 < 690,212
   85,360 = 85,360

8. Trenton, Toms River, Elizabeth, Paterson, Jersey City, Newark

9. A. 19,304 > 19,298
   B. 25,189 < 25,190
   C. 32,400 > 32,375
   D. 57,626 = 57,626

10. A. N
    B. Y
    C. N
    D. N
    E. Y
    F. Y

11. No. The largest number that rounds to 36,000 is 36,499.

12. A. 9,300
    B. 7,205
    C. 6,205
    D. 1,359

13. 538 + 774 = 1,312

14. Nicole did not solve the problem correctly. She did not regroup to subtract.

726	regroups to	600 + 110 + 16
	subtracts	500 + 40 + 9
	equals	100 + 70 + 7 = 177

15. (6 × 3,000) + (6 × 100) + (6 × 40) + (6 × 2)

16. **B**, ***C***, and **E**

17. 672
    5,376 ÷ 8 = 672

18. 3,038
    4 × 759 = 3,036 + 2 = 3,038

Chapter Checklist for Standards

If you answered the questions correctly, you are on your way toward mastering the concepts and skills for this standard.

Standard	Concept/Skill	Questions
4.NBT.1	I can recognize that in multidigit whole numbers, the value of a digit increases 10 times when the digit moves one place to the left.	5
4.NBT.2	I can read and write whole numbers using standard form, word form, and expanded form, and compare/order pairs and sets of numbers.	1, 2, 3, 4, 7
4.NBT.3	I can round numbers to any place.	6, 8, 9, 10, 11
4.NBT.4	I can fluently add and subtract large numbers.	12, 13, 14
4.NBT.5	I can multiply whole numbers using strategies based on place value, equations, and models.	12, 15, 17, 18
4.NBT.6	I can find whole-number quotients using strategies based on place value, equations, and models.	12, 16, 17, 18

Numbers and Operations—Fractions (NF)

CHAPTER 5

VOCABULARY

Fractions: values that represent

- Equal parts of a whole
- Locations or points on a number line with equal distances between whole numbers
- Equal parts of a set

Numerator: number above the fraction bar that names how many equal parts

Denominator: number below the fraction bar that names how many equal parts in the whole or whole set

Fraction bar: located below the numerator and above the denominator

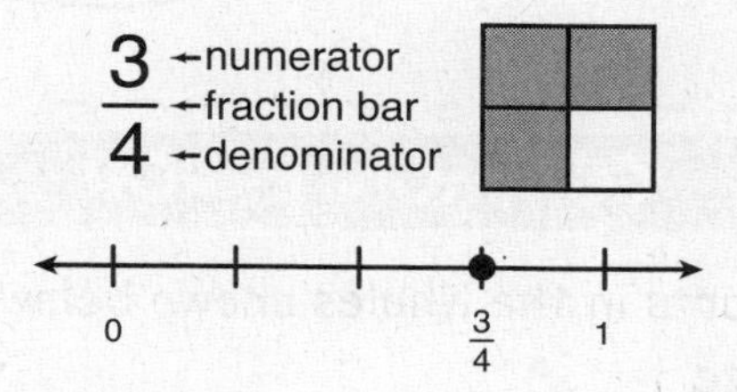

Unit fraction: fraction with a numerator of 1

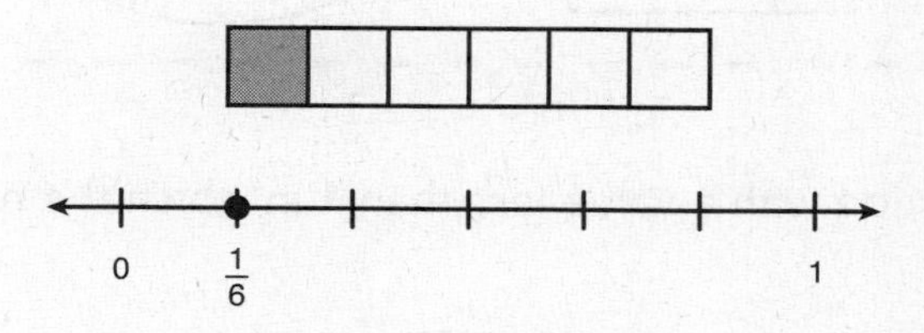

COMMON ERROR

There are 7 total tic marks including the 0 and the 1 but that is **not** the number of equal parts or distances, so it is not the denominator. The denominator is the number of equal spaces between 0 and 1, which is 6 spaces, or sixths. The point is located at the first equal space, or $\frac{1}{6}$.

Fractions as Equal Parts

Equal Parts of a Whole: Correct Fractions

1 part shaded out of 4 total parts = $\frac{1}{4}$ or "one-fourth"

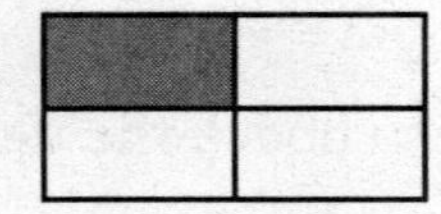

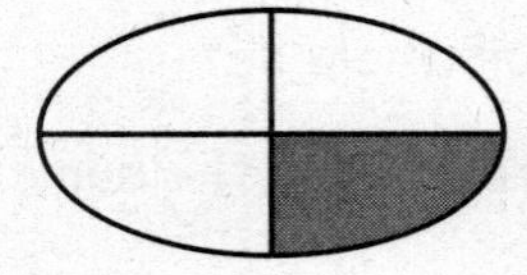

COMMON ERROR

There are 4 parts in the wholes shown below, but they are **not** equal in size.

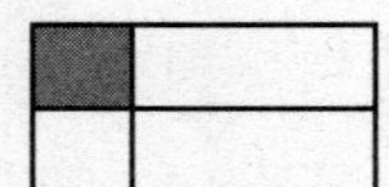

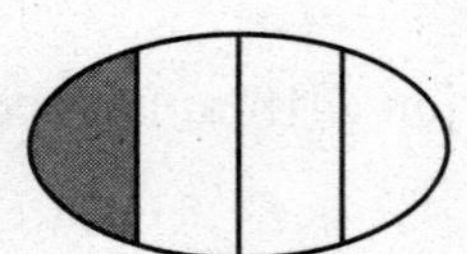

Proper fraction: a fraction with a value less than 1 in which the numerator is less than the denominator

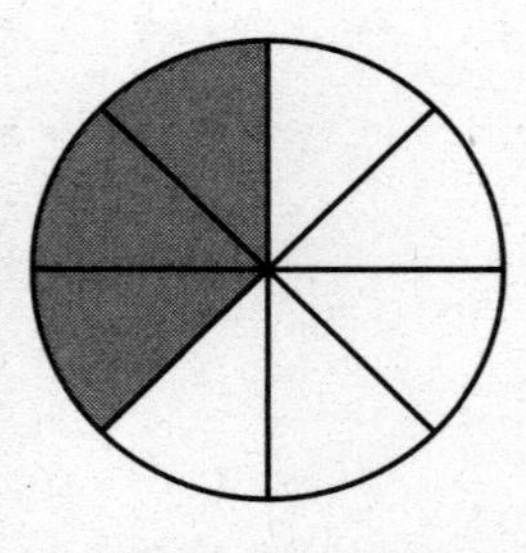

$\frac{3}{8}$ Proper Fraction

Equal Parts of a Set

There is a set of 8 circles. In the set, 3 of the circles are shaded and 5 are unshaded.

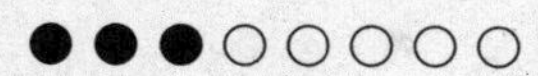

The fraction of the set that is shaded = $\frac{3}{8}$ (three-eighths)

The fraction of the set that is not shaded = $\frac{5}{8}$ (five-eighths)

> **COMMON ERROR**
>
> The denominator must be the total number of circles in the set, **not** the number of circles that are unshaded.

Fractions Equal to One Whole

When the numerator and the denominator are the same, the fraction equals 1.

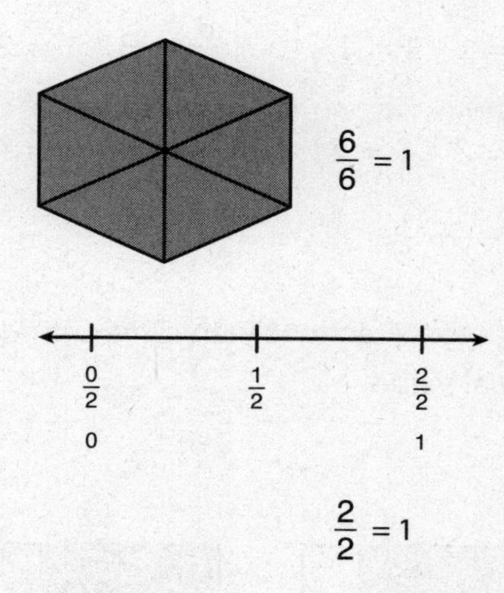

Fractions Equal to Whole Numbers

We can express any number as a fraction, even whole numbers.

Each whole has only 1 part, so the denominator is 1. There are 3 wholes, so the numerator is 3.

$$3 \text{ wholes} = \frac{3}{1}$$

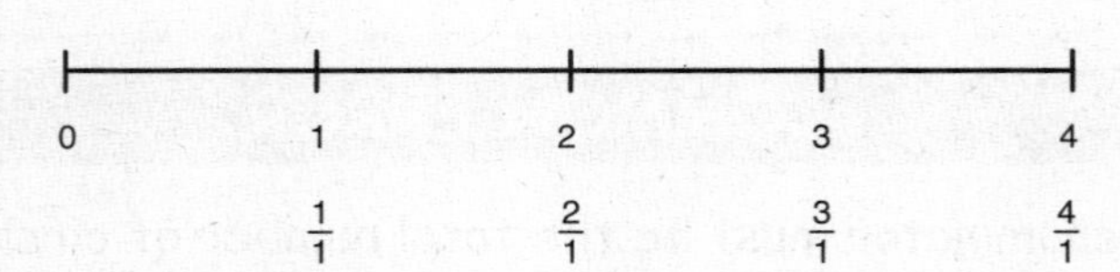

There is one space between whole numbers.

WHY DOES THIS WORK?

The fraction bar is actually a division sign. The value of the fraction is the quotient of the numerator and denominator:

$$\frac{1}{1} = 1 \div 1 = 1 \qquad \frac{3}{1} = 3 \div 1 = 3$$

$$\frac{2}{1} = 2 \div 1 = 2 \qquad \frac{4}{1} = 4 \div 1 = 4$$

Equivalent fractions: two or more fractions with different numerators and denominators that have equal value.

Equivalent shaded amounts:

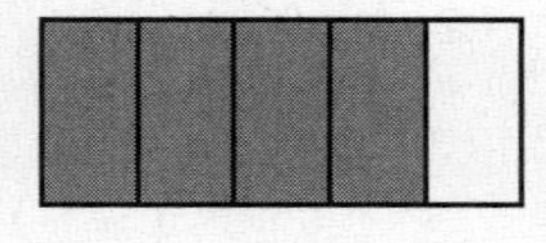 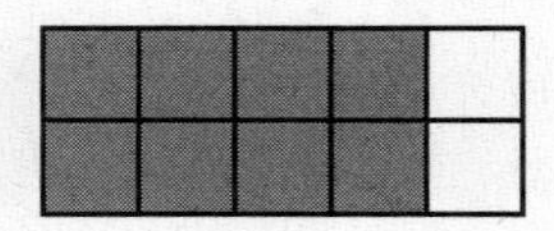

$$\frac{4}{5} = \frac{8}{10}$$

The shaded amounts in each shape are the same, so the fractions are equivalent.

Equivalent Points on a Number Line

The points are located at the same part on the number line between 0 and 1.

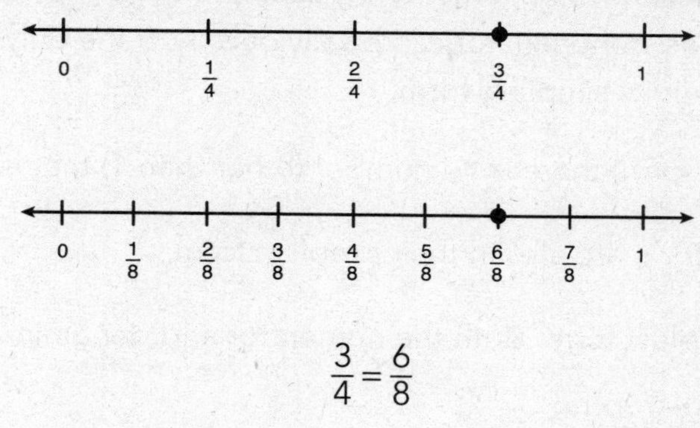

$$\frac{3}{4} = \frac{6}{8}$$

Finding Equivalent Fractions

Equivalent fractions can be found by using shaded shapes and number lines. They can also be found by using multiplication strategies:

$$\frac{2}{3} = \frac{?}{6}$$

Strategy

$3 \times ? = 6$

Use the same number to multiply both parts of the fraction:

$$\frac{2 \times 2}{3 \times 2} = \frac{4}{6}$$

$$\frac{1}{6} = \frac{2}{?} \qquad \frac{1 \times 2}{6 \times 2} = \frac{2}{12}$$

WHY DOES THIS WORK?

The identity property of multiplication tells us that when a number is multiplied by 1, the number has the same value:

$$3 \times 1 = 3$$

When you multiply a fraction by a fraction equal to 1, the original fraction and the product will have the same value:

$$\frac{3}{5} \times \frac{2}{2} = \frac{6}{10}$$

Renaming Fractions in Simplest Terms

We can find the simplest form of a fraction by seeing if there is a common factor that both the numerator and denominator can be divided by. If the only factor is 1, then the fraction is already in its **simplest form.**

$\frac{1}{3}$ is in its simplest form; there is no number (other than 1) that is a common factor of both digits. $\frac{4}{5}$ and $\frac{5}{12}$ are also in their simplest form.

$\frac{3}{12}$ is not in its simplest form. Both the numerator and denominator can be divided by 3 so there is a simpler form:

$$\frac{3 \div 3}{12 \div 3} = \frac{1}{4}$$

Strategy

To simplify a fraction, find the greatest common factor (GCF) for the numerator and denominator, then divide by making the factor a fraction equal to 1 whole. When dividing by a fraction equal to 1 whole, the value remains the same based on the identity property of multiplication and division.

$\frac{8}{10}$ Factors of 8: 1, [2], 4, 8

Factors of 10: 1, [2], 5, 10

The greatest common factor (GCF) is 2:

$$\frac{8 \div 2}{10 \div 2} = \frac{4}{5}$$

$\frac{6}{12}$ Factors of 6: 1, 2, 3, [6]

Factors of 12: 1, 2, 3, 4, [6], 12

The greatest common factor (GCF) is 6:

$$\frac{6 \div 6}{12 \div 6} = \frac{1}{2}$$

Know your fractions equal to $\frac{1}{2}$ to quickly find the simplest form.

COMMON ERROR

Finding a simpler form but not seeing that the fraction can be further simplified.

$\frac{8}{12}$ can be simplified to $\frac{4}{6}$, but it is not the simplest form.

The fraction can be further simplified to $\frac{2}{3}$:

$$\frac{8 \div 2}{12 \div 2} = \frac{4}{6}; \frac{4 \div 2}{6 \div 2} = \frac{2}{3}$$

Improper Fractions and Mixed Numbers

Fractions that represent whole numbers and parts of a whole can be expressed as improper fractions and mixed numbers.

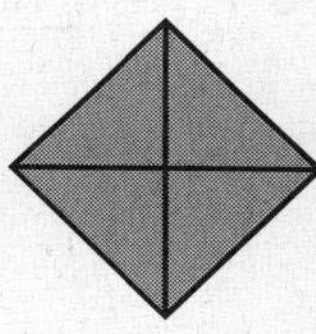

$\frac{4}{4}$, or 1 whole

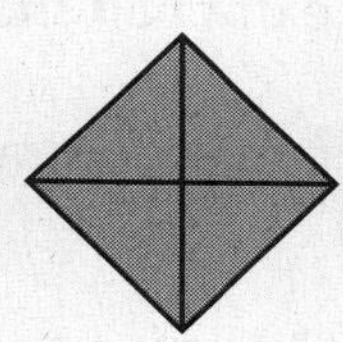

$\frac{4}{4}$, or 1 whole

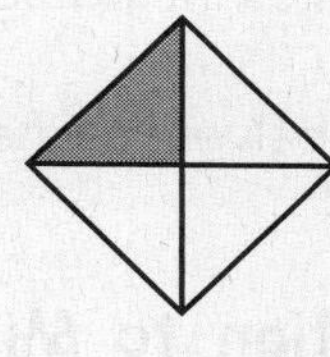

$\frac{1}{4}$

Mixed number: counts the whole numbers and a fraction: $2\frac{1}{4}$

Improper fraction: counts only the number of parts (fourths): $\frac{9}{4}$

The numerator is larger than the denominator because the value is greater than 1.

Renaming Improper and Mixed Fractions

We can name the same fraction with both improper and mixed numbers.

Mixed number to improper fraction: $3\frac{1}{2} = \frac{?}{2}$

Visual Strategy

Show the fractions for the shaded amounts for each whole number and part of a whole.

Combine the fractions:

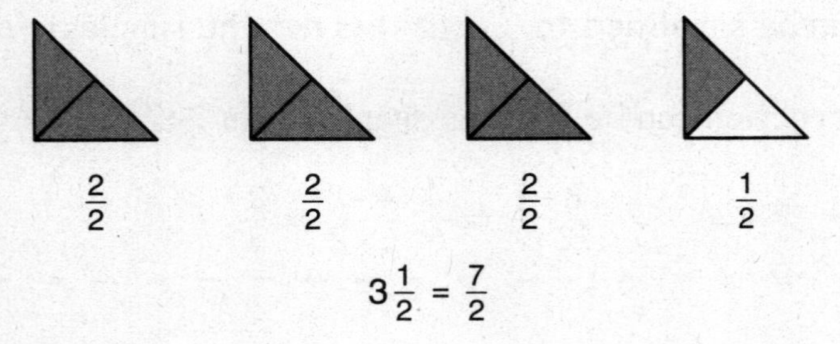

Reasoning Strategy

How many halves in one whole? <u>2 halves</u>

How many halves in three wholes? <u>6 halves</u>

How many total halves including the extra half shown? $6 + 1 =$ <u>7 halves</u>

Combine all of the halves to find your answer. How many halves in $3\frac{1}{2}$? <u>7 halves</u>

Improper Fraction to Mixed Number

$$\frac{23}{5} = ?\frac{?}{5}$$

Visual Strategy

Show the shaded amounts for all of the wholes and parts:

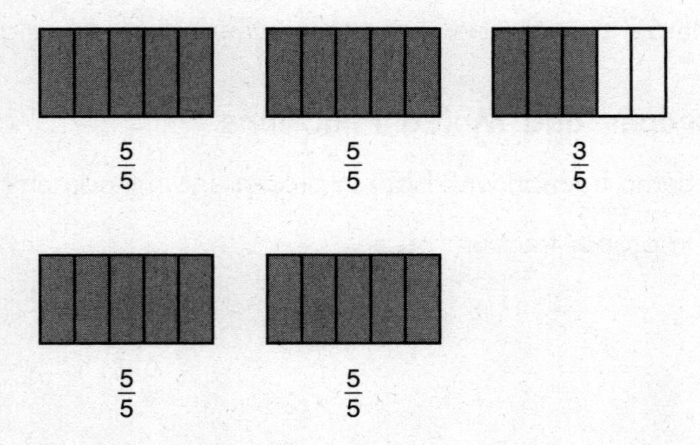

Reasoning Strategy

How many wholes $\left(\frac{5}{5}\right)$ in $\frac{23}{5}$? There are 4, which totals $\frac{20}{5}$.

What fraction is left? $\frac{3}{5}$

Combine both fractions to find your answer: $\frac{23}{5} = 4\frac{3}{5}$

Comparing Fractions

We can compare the values of **like fractions** (fractions with the same denominator).

Visual Strategy

Compare the shaded amounts:

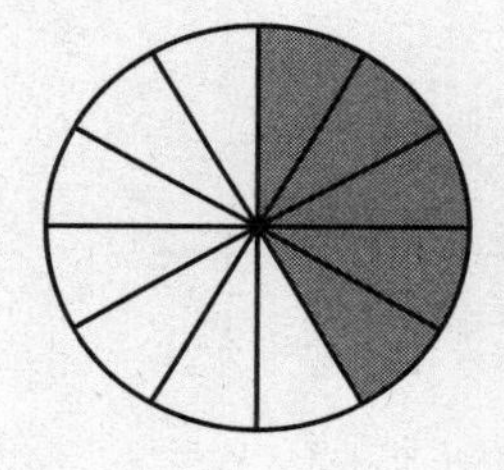

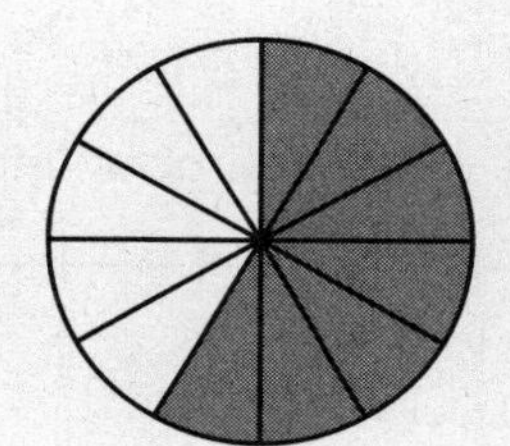

$\frac{5}{12}$ is less than $\frac{7}{12}$

$\frac{5}{12} < \frac{7}{12}$. The symbol for less than is $<$.

Visual Strategy

Compare the locations of points on a number line:

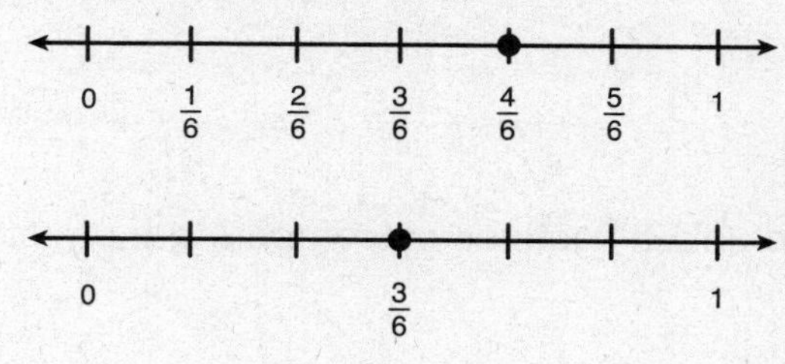

The point at $\frac{4}{6}$ is closer to the whole number 1 than the point at $\frac{3}{6}$.

$\frac{4}{6}$ is greater than $\frac{3}{6}$.

$\frac{4}{6} > \frac{3}{6}$. The symbol for greater than is $>$.

We can compare the values of fractions with the same numerators and different denominators.

Visual Strategy

Compare the shaded amounts:

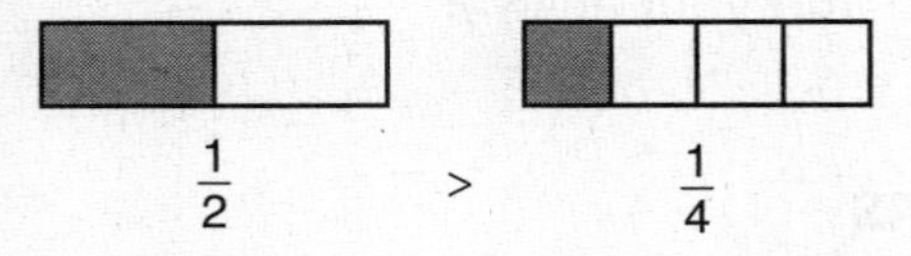

1 out of 2 parts is greater than 1 out of 4 parts.

We can compare the values of **unlike fractions** (fractions with different numerators and denominators).

Visual Strategy

Compare the locations on the number lines:

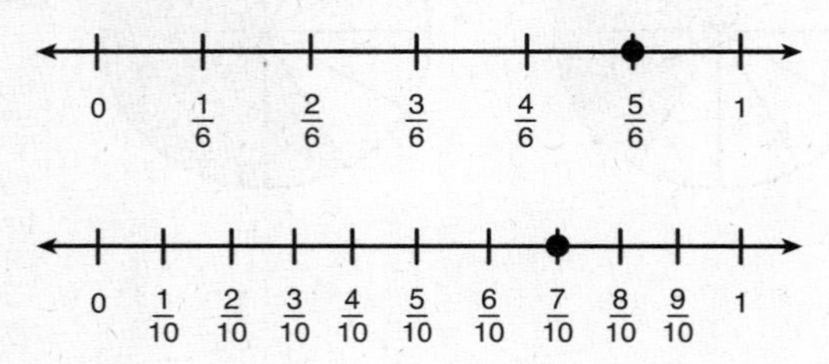

The point at $\frac{5}{6}$ is closer to one whole than the point at $\frac{7}{10}$.

$$\frac{5}{6} > \frac{7}{10}$$

Reasoning Strategy

Use benchmarks.

Sometimes we can compare unlike fractions by comparing them to 0, $\frac{1}{2}$, and 1.

Know all of your fractions equivalent to $\frac{1}{2}$:

$$\frac{1}{2} = \frac{2}{4} = \frac{3}{6} = \frac{4}{8} = \frac{5}{10} = \frac{6}{12}$$

Compare: $\frac{3}{10}$? $\frac{5}{6}$

$\frac{3}{10}$ is less than half. $\frac{5}{6}$ is greater than half. $\frac{3}{10} < \frac{5}{6}$

Compare: $\frac{4}{3}$? $\frac{11}{12}$

$\frac{4}{3}$ is greater than one whole. $\frac{11}{12}$ is less than one whole. $\frac{4}{3} > \frac{11}{12}$

COMMON ERROR

Comparing fractions by comparing the numbers.

$\frac{11}{12}$ has larger digits than $\frac{4}{3}$, but $\frac{4}{3}$ is larger because it is greater than 1 whole.

Additional Strategy

Find common denominators.

Compare: $\frac{3}{4}$? $\frac{2}{3}$

Find the common denominator by finding the first number that is a multiple of both 4 and 3. We call this the **least common multiple (LCM)**.

Strategy

Make a list.

Multiples of 3: 3, 6, 9, [12,] 15

Multiples of 4: 4, 8, [12,] 16

LCM = 12

$$\frac{3}{4}\times\frac{3}{3}=\frac{9}{12} \qquad \frac{2}{3}\times\frac{4}{4}=\frac{8}{12} \qquad \frac{9}{12}>\frac{8}{12} \text{ so } \frac{3}{4}>\frac{2}{3}$$

Adding and Subtracting Like Fractions

We can add like fractions by adding the numerators.

Addition

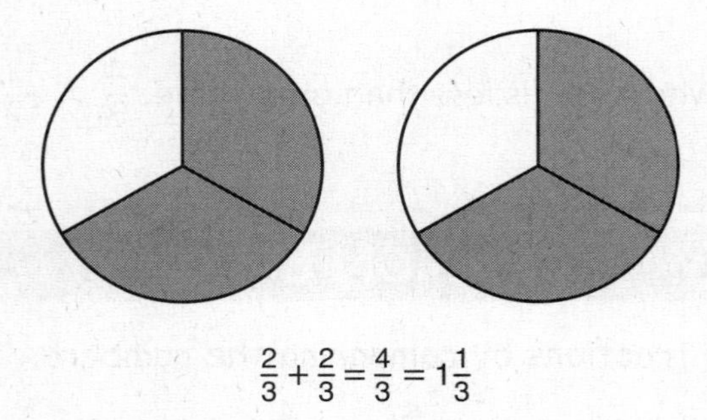

$$\frac{2}{3}+\frac{2}{3}=\frac{4}{3}=1\frac{1}{3}$$

COMMON ERROR

Add only the numerators, not the denominators.

Why?

When you add 2 pencils + 2 pencils, you get 4 pencils.

When you add fractions, think of the denominators the same as you think of objects.

2 thirds + 2 thirds = 4 thirds

Subtraction

$$\frac{7}{8}-\frac{4}{8}=\frac{3}{8}$$

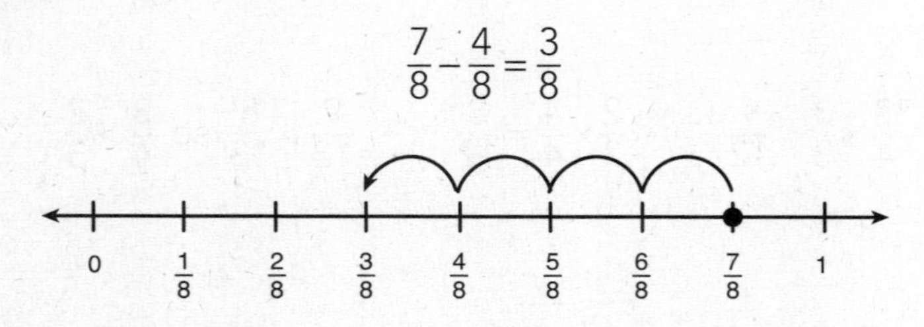

Using Repeated Addition to Multiply Whole Numbers and Fractions

We know that we can multiply whole numbers by using repeated addition:

$$5 + 5 + 5 + 5 + 5 + 5 = 6 \times 5 = 30$$

We can multiply a fraction by using repeated addition:

$$\frac{1}{5}+\frac{1}{5}+\frac{1}{5}+\frac{1}{5}+\frac{1}{5}+\frac{1}{5}=6\times\frac{1}{5}=\frac{6}{5}=1\frac{1}{5}$$

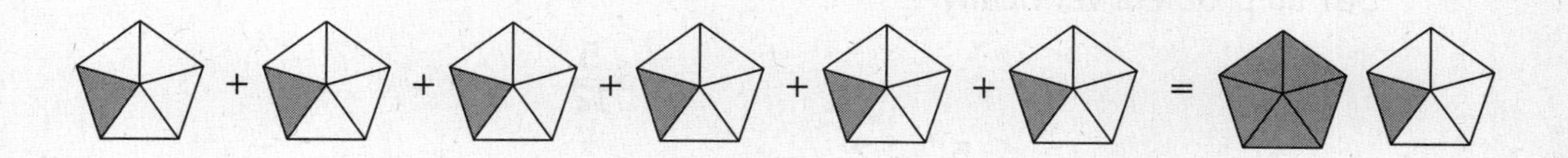

$$\frac{2}{8}+\frac{2}{8}+\frac{2}{8}+\frac{2}{8}=4\times\frac{2}{8}=\frac{8}{8}=1$$

$0 \quad \frac{1}{8} \quad \frac{2}{8} \quad \frac{3}{8} \quad \frac{4}{8} \quad \frac{5}{8} \quad \frac{6}{8} \quad \frac{7}{8} \quad 1$

Note

$4\times\frac{2}{8}=8\times\frac{1}{8}$. Both equations are equal to 1 whole.

Adding Mixed Numbers

Strategy

First, add or subtract the fractions, then the whole numbers. Check to see if the sum has an improper fraction that needs to be renamed.

Set up problem vertically:

$$3\frac{5}{12}+2\frac{2}{12}=?\qquad \begin{array}{r} 3\frac{5}{12} \\ +2\frac{2}{12} \\ \hline 5\frac{7}{12} \end{array}$$

For addition, check to see if the answer is an improper fraction that should be renamed as a mixed number:

$$4\frac{5}{6}+1\frac{2}{6}=?\qquad \begin{array}{r} 4\frac{5}{6} \\ +1\frac{2}{6} \\ \hline 5\frac{7}{6} \end{array}$$

The sum of $5\frac{7}{6}$ can be renamed. First, rename the improper fraction $\frac{7}{6}$ to $1\frac{1}{6}$.

Next, add the whole number 5 to the mixed number: $5 + 1\frac{1}{6} = 6\frac{1}{6}$.

COMMON ERROR

Don't forget to add the whole number to the mixed number in the sum.

Subtracting Mixed Numbers

Strategy

Check to see if the whole number has to be regrouped before subtracting.

Set up problem vertically:

$$1\frac{1}{10}-\frac{3}{10}=?$$

$$1\frac{1}{10}=\frac{10}{10}+\frac{1}{10}=\frac{11}{10}$$

$$-\frac{3}{10}$$

$$\begin{array}{r} \frac{11}{10} \\ -\frac{3}{10} \\ \hline \frac{8}{10} \end{array}$$

You cannot subtract $\frac{1}{10}-\frac{3}{10}$, so rename $1\frac{1}{10}$ as the improper fraction $\frac{11}{10}$ before subtracting.

Fractions and Decimals

We can rename fractions with numerators of 10 and 100 as decimals.

Place value of whole numbers and decimals:

____	____	•	____	____
Tens	Ones	Decimal Point	Tenths	Hundredths

The decimal point is placed between the whole number and the decimal.

The chart below shows fractions and their equivalent decimals:

$\frac{1}{10}$	$\frac{1}{10}$	$\frac{1}{10}$	$\frac{1}{10}$	$\frac{1}{10}$	$\frac{1}{10}$	$\frac{1}{10}$	$\frac{1}{10}$	$\frac{1}{10}$	$\frac{1}{10}$
0.1	0.1	0.1	0.1	0.1	0.1	0.1	0.1	0.1	0.1

In the model, the whole has 10 equal parts. Each part is $\frac{1}{10}$, or 0.1 of the whole.

For decimals less than 1, a "0" can be placed in the ones place.

For decimals more than 1, the value of the ones is placed in the ones place.

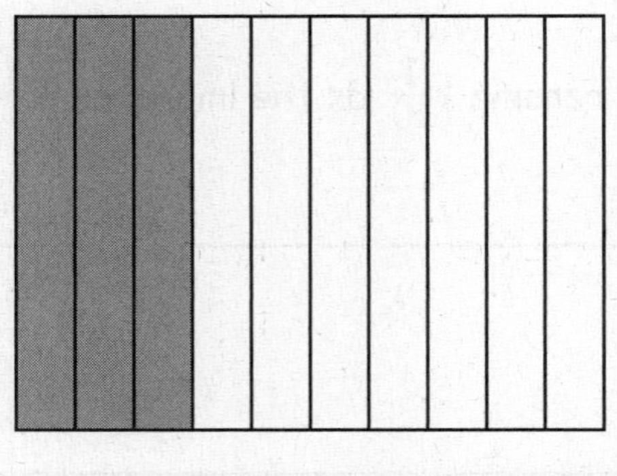

$\frac{3}{10}$, or 0.3 of the whole is shaded.

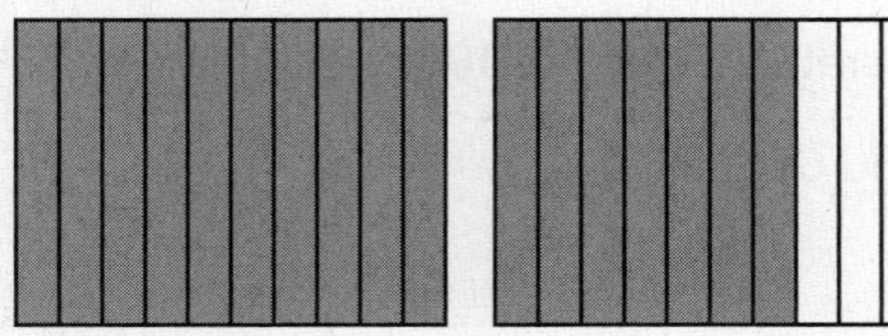

$1\frac{7}{10}$, or 1.7 is shaded.

We express fractions and decimals using the same words.

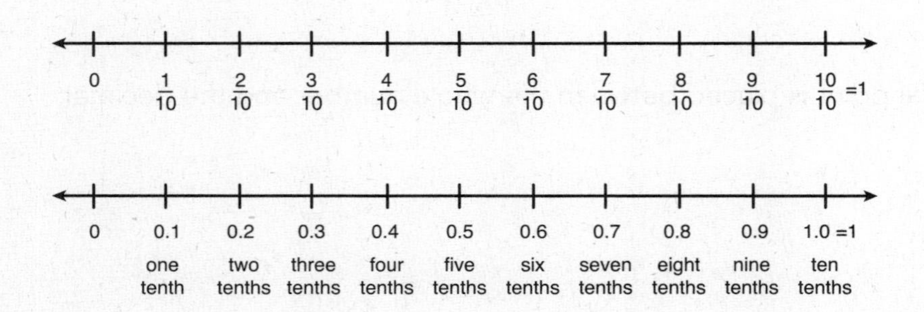

WHY DOES THIS WORK?

Why is 0.1 more than 0.9 equal to 1.0?

Think about adding 0.1 to each number when you move to the right on the number line.

$$\begin{array}{r} 0.1 \\ +\,0.1 \\ \hline 0.2 \end{array} \qquad \begin{array}{r} 0.5 \\ +\,0.1 \\ \hline 0.6 \end{array} \qquad \begin{array}{r} {}^{1}0.9 \\ +\,0.1 \\ \hline 1.0 \end{array}$$

We add 9 + 1 in the tenths place and carry the 1 to the ones place.

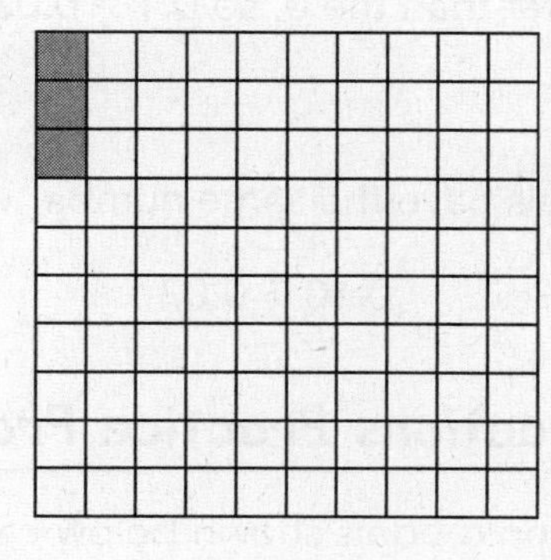

$\frac{3}{100}$, or 0.03 is shaded.

COMMON ERROR

Remember to place a 0 in the tenths place. The shaded amount above is 0.03, not 0.3.

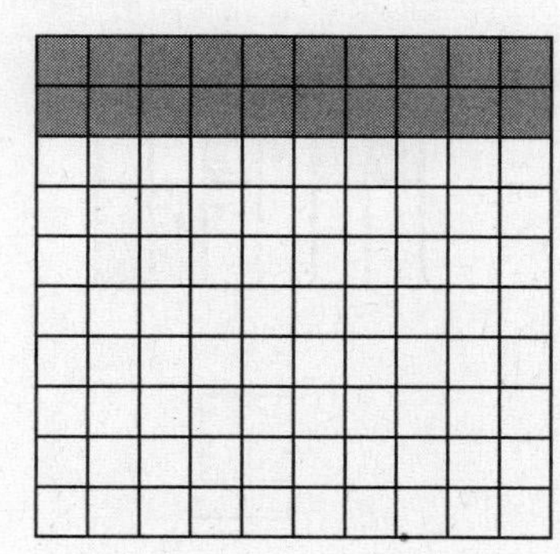

$\frac{20}{100}$, or 0.20 is shaded.

Also $\frac{2}{10}$, or 0.2 is shaded.

Note: Ten hundredths is equal to one tenth.

Comparing Decimals

Strategy 1

Vertically stack the decimals to compare the digits in each place.

Compare 0.1 to 0.03

Ones	Tenths	Hundredths
0.	1	
0.	0	3

Notice in the tenths place, the 1 is greater than the 0, so $0.1 > 0.03$.

Strategy 2

Add 0 to decimal places so both decimals have the same number of places:

$0.1 = 0.10$ $\qquad$ $0.10 > 0.03$

Numbers and Operations—Fractions Practice Problems

1. The shaded amounts of the fraction models shown below can be used to form an equation. Place the numbers and symbols in the boxes shown to make a correct equation for the fraction model.

   Numbers/symbols to use:

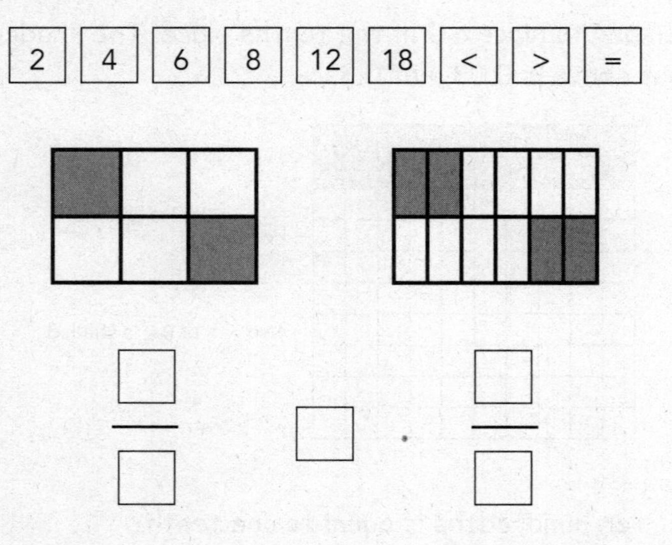

2. Ethan has a bottle of juice. The amount of juice, in liters, is shown in the illustration to the right.

Ethan poured some juice from the bottle into a glass. After he poured the juice into the glass, there was some juice left in the bottle. What could be the amount of juice, in liters, that Ethan poured into the glass? Choose THREE that could be correct.

☐ A. $\frac{5}{12}$ liter

☐ B. $\frac{3}{4}$ liter

☐ C. $\frac{3}{10}$ liter

☐ D. $\frac{1}{3}$ liter

☐ E. $\frac{4}{6}$ liter

☐ F. $\frac{1}{2}$ liter

3. Sophia has 1 yard of ribbon. After she cut a piece of ribbon, she has $\frac{7}{12}$ yard left.

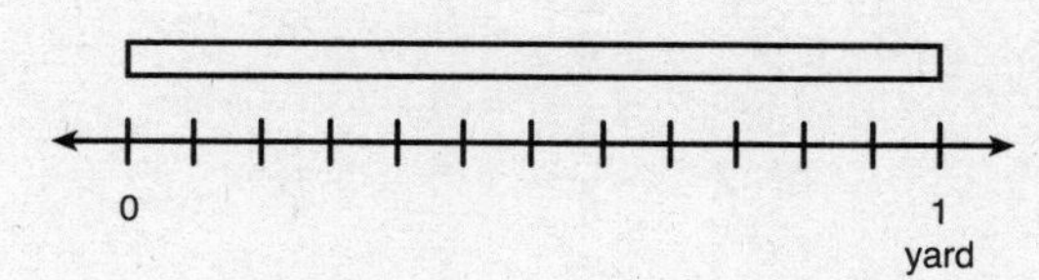

PART A

What was the total amount of ribbon, in yards, that Sophia cut? Write an equation to help you find your answer.

PART B

Solve the equation by writing your answer in the sentence below:

Sophia cut $\frac{\square}{\square}$ yards of ribbon.

4. Look at the number line below:

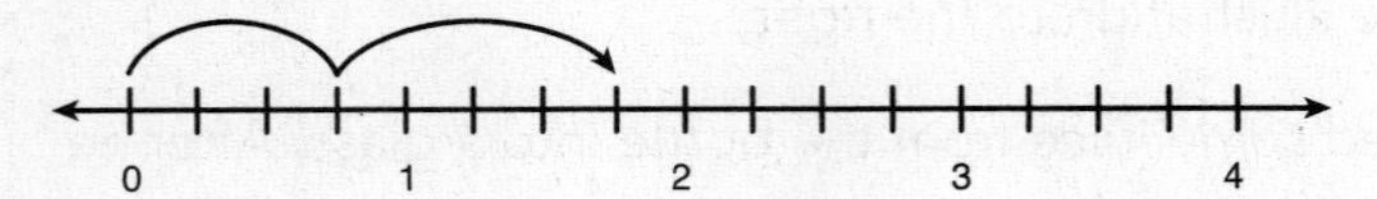

Which equation is modeled on the number line?

- ○ A. $\frac{3}{4}+\frac{4}{4}=\frac{7}{4}$
- ○ B. $\frac{4}{5}+\frac{4}{5}=\frac{8}{5}$
- ○ C. $\frac{3}{7}+\frac{4}{7}=\frac{7}{7}$
- ○ D. $\frac{3}{8}+\frac{4}{8}=\frac{7}{8}$

5. The shaded amounts of the fraction model shown below can be used to solve the equation below:

$$\frac{2}{6} + \frac{2}{6} = \frac{4}{6}$$

$$\text{or } 2 \times \frac{2}{6} = \frac{4}{6}$$

PART A

Shade the shapes below to show another way to find the same sum. Use all of the shapes shown.

+ + + =

PART B

Write a new equation that can be used with the shaded shapes above.

6. Look at the fraction model below:

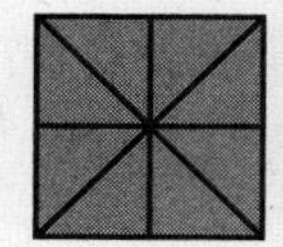 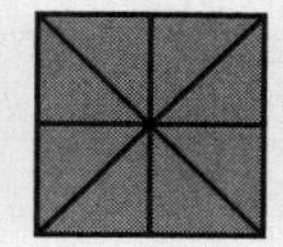 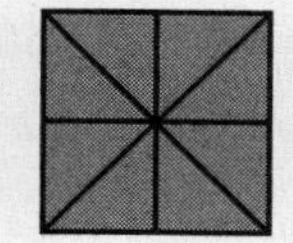 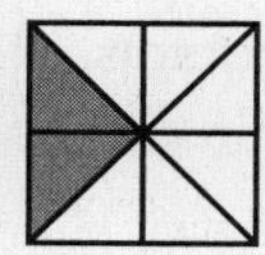

What is the shaded amount shown? Choose THREE that are correct.

☐ A. $3\frac{1}{4}$

☐ B. $3\frac{2}{6}$

☐ C. $3\frac{2}{8}$

☐ D. $\frac{26}{8}$

☐ E. $\frac{31}{4}$

☐ F. $\frac{32}{6}$

7. The chart below shows the heights of some of the fourth-grade students when they were in kindergarten.

Student	Height in Kindergarten (Feet)
Kiera	$3\frac{6}{12}$
Mario	$3\frac{7}{12}$
Tyrell	$3\frac{5}{12}$
Zoey	$3\frac{3}{12}$

PART A

Since kindergarten, Tyrell grew $1\frac{6}{12}$ feet and Zoey grew $1\frac{9}{12}$ feet.

Tyrell said that he is now taller than Zoey. Is Tyrell correct? Why or why not?

PART B

Mario's height in fourth grade is now $4\frac{10}{12}$ feet. How much did Mario grow since kindergarten? Write an equation to find your answer.

Solve your equation.

8. Justin has 8 baseballs, each with a weight of $\frac{1}{5}$ kilogram. What is the total weight, in kilograms, of the 8 baseballs? Choose and place the numbers shown in the boxes to show an equation that can be used to find the total weight.

Numbers/symbols to use:

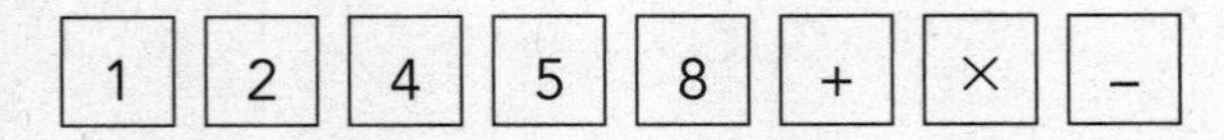

(You can use some numbers more than once.)

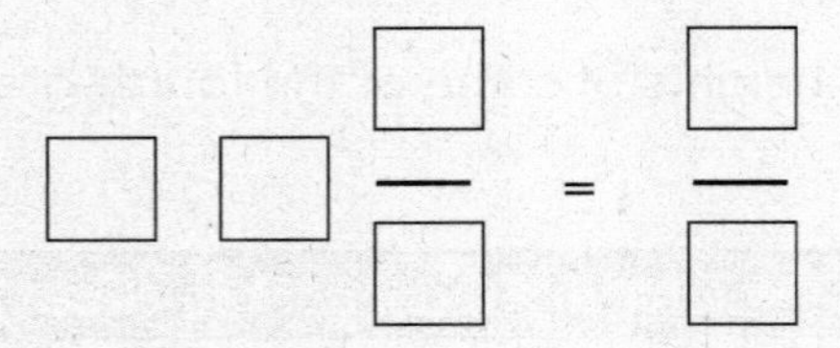

9. On the number lines shown, point *A* represents the fraction $\frac{2}{3}$. Point *B* represents a fraction that is equivalent to $\frac{2}{3}$.

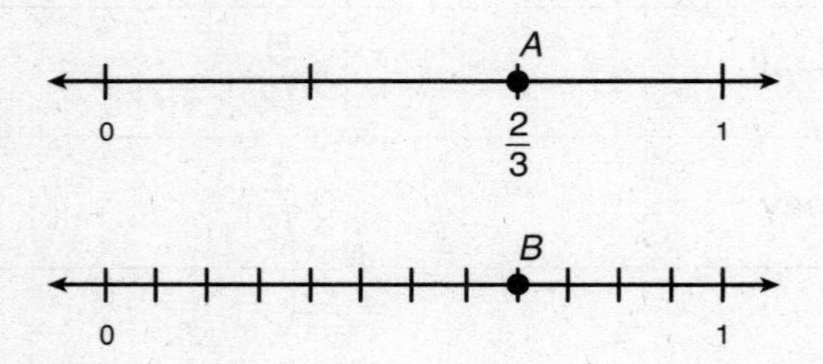

Use the numbers shown to construct an equation that can be used to find the fraction represented by point *B* that is equivalent to point *A*:

Numbers to use:

4	6	8	9	12

(You can use some numbers more than once.)

$$\frac{2 \times \square}{3 \times \square} = \frac{\square}{\square}$$

10. Place the fractions shown in order from the least to the greatest.

$\frac{1}{3}$ $\frac{1}{6}$ $1\frac{1}{12}$ $\frac{1}{1}$ $\frac{1}{8}$ $1\frac{1}{5}$

______ ______ ______ ______ ______ ______

least greatest

11. The shaded amounts represent which of the following values? Choose THREE that are correct.

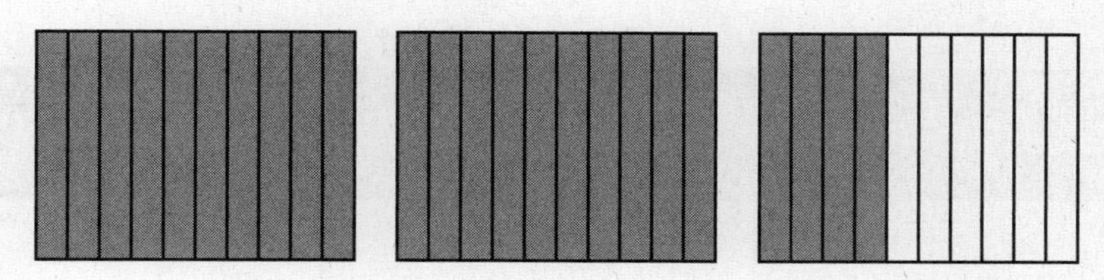

☐ A. $2\frac{4}{10}$ ☐ C. $\frac{24}{10}$ ☐ E. 2.4

☐ B. $2\frac{4}{100}$ ☐ D. $\frac{24}{100}$ ☐ F. 2.04

12. Jake is on the track team. The results of the 100-meter race are shown on the chart below:

Runner	Race Time (Seconds)
Jake	13.5
Conor	13.07
Dimar	13.2
Matt	13.09

The winner of the race is the runner with the fastest time, or least decimal value. Which runner was the winner of the race?

O A. Jake
O B. Conor
O C. Dimar
O D. Matt

13. Jarrad wrote down some fractions and decimals. He wants to place them in the chart shown based on their value compared with $\frac{1}{2}$ and 1.

Place the fractions and decimals below in Jarrad's table to correctly show their value compared with the values shown.

$\frac{1}{4}$	0.6	$1\frac{1}{3}$	0.09	$\frac{7}{8}$	$\frac{3}{2}$	1.01	0.44

Jarrad's Chart

Values < $\frac{1}{2}$(0.5)	$\frac{1}{2}$ < Value < 1	Values > 1

Answers to Numbers and Operations—Fractions Practice Problems

1. $\frac{2}{6}=\frac{4}{12}$. In the model on the left, 2 out of 6 parts are shaded and on the right, 4 out of 12 parts are shaded. For both models, the same fractional amount is shaded, so the fractions are equivalent. You can also use multiplication to find equivalent fractions.

$$\frac{2}{6}\times\frac{2}{2}=\frac{4}{12}$$

When multiplying a fraction by a value equal to 1 whole, the result is an equivalent fraction.

2. **A, C, D** The bottle is $\frac{1}{2}$ full, so if Ethan poured some juice and left some in the bottle, the amount poured has to be less than $\frac{1}{2}$.

A Correct: $\frac{6}{12} = \frac{1}{2}$, so $\frac{5}{12} < \frac{1}{2}$

B. Incorrect: $\frac{2}{4} = \frac{1}{2}$, so $\frac{3}{4} > \frac{1}{2}$.

C Correct: $\frac{5}{10} = \frac{1}{2}$, so $\frac{3}{10} < \frac{1}{2}$

D Correct: $\frac{1}{3} < \frac{1}{2}$; 1 out of 3 parts is less than 1 out of 2 parts.

E. Incorrect: $\frac{3}{6} = \frac{1}{2}$, so $\frac{4}{6} > \frac{1}{2}$

F. Incorrect: $\frac{1}{2} = \frac{1}{2}$, so pouring $\frac{1}{2}$ will not leave any amount in the bottle.

3. Part A: Equation: $\frac{12}{12} - ? = \frac{7}{12}$ or $\frac{12}{12} - \frac{7}{12} = ?$

The total ribbon has a length of $\frac{12}{12}$ yard. If there is $\frac{7}{12}$ yard left, you can find the amount cut by subtracting.

Part B: Solution: Sophia cut $\frac{5}{12}$ yards of ribbon.

4. **A** $\frac{3}{4}+\frac{4}{4}=\frac{7}{4}$. The number line has 4 equal spaces or parts between 0 and 1, so the denominator is 4. The model represents starting with a value of 3 parts and adding a value of 4 more parts. Therefore, A is correct.

5. In the model, 2 parts out of 6 are shaded in two wholes. The model also shows the sum as 4 out of 6 shaded parts.
The second model has 4 wholes and has to have a sum of 4 out of 6 parts.

If each of the 4 wholes has 1 shaded part, then the sum will also be 4 out of 6 parts.

Part A:

Part B: Equation: $\frac{1}{6} + \frac{1}{6} + \frac{1}{6} + \frac{1}{6} = \frac{4}{6}$ or $4 \times \frac{1}{6} = \frac{4}{6}$

6. **A, C, D** There are 8 equal parts in each whole, so the denominator is 8. The model shows 3 wholes and 2 out of 8 parts are shaded in the last whole.

   The model shows $\frac{8}{8}$ shaded in each of 3 wholes and $\frac{2}{8}$ shaded in the last whole.

   The amount shaded is $3\frac{2}{8}$ or an equivalent mixed number or improper fraction.

   A Correct: $\frac{2}{8} = \frac{1}{4}$, so $3\frac{2}{8} = 3\frac{1}{4}$

   C Correct: $1 + 1 + 1 + \frac{2}{8} = 3\frac{2}{8}$

   D Correct: $\frac{8}{8} + \frac{8}{8} + \frac{8}{8} + \frac{2}{8} = \frac{26}{8}$

7. For Tyrell's height add: $3\frac{5}{12} + 1\frac{6}{12} = 4\frac{11}{12}$

   For Zoey's height add: $3\frac{3}{12} + 1\frac{9}{12} = 4\frac{12}{12} = 5$

   Part A: Tyrell is not correct. Tyrell is now $4\frac{11}{12}$ feet and Zoey is now $4\frac{12}{12}$ feet (or 5 feet).

   Part B: Equation: $3\frac{7}{12} + ? = 4\frac{10}{12}$ or $4\frac{10}{12} - 3\frac{7}{12} = ?$

   We know Mario's new height and his old height. In order to find out how much he grew, we need to subtract the new height – old height.

   Solution: $4\frac{10}{12} - 3\frac{7}{12} = 1\frac{3}{12}$

8. Equation: $8 \times \frac{1}{5} = \frac{8}{5}$

9. In the first number line, point *A* is located at $\frac{2}{3}$.

   In the second number line, point *B* is located at the same distance from 0 as point *A* so the fractions are equivalent.

Multiplying the given value of $\frac{2}{3}$ by a fraction equal to 1 whole will result in an equivalent fraction.

We know the denominator is 12 because there are 12 equal parts between 0 and 1 on the second number line. Set up the equation with the known part of the second fraction.

$\frac{2}{3} = \frac{?}{12}$ Ask: What do you multiply 3 by to get 12? $\frac{2}{3} \times \frac{}{4} = \frac{?}{12}$

Now multiply by a fraction equal to 1 whole: $\frac{2}{3} \times \frac{4}{4} = \frac{?}{12}$

Solution: $\frac{2}{3} \times \frac{4}{4} = \frac{8}{12}$

10. Correct order: $\frac{1}{8}, \frac{1}{6}, \frac{1}{3}, \frac{1}{1}, 1\frac{1}{12}, 1\frac{1}{5}$

Since all of the numerators are the same, we order the fractions by the denominator. The smaller the denominator, the larger the fraction. Mixed numbers are greater than 1.

11. **A, C, E** Each of the 2 shaded wholes has 10 equal parts so the denominator is 10 for a fraction and the decimal is in tenths. In the last whole, 4 out of 10 parts are shaded.

The shaded amount is equal to 2 wholes and 4 out of 10.

A Correct: $1 + 1 + \frac{4}{10} = 2\frac{4}{10}$

C Correct: $2\frac{4}{10} = \frac{24}{10}$

E Correct: $2\frac{4}{10} = 2.4$

12. **B** The winner has the lowest time. Stack the numbers to compare place values. There are 2 decimals with a 0 in the tenths place, so continue to the hundredths place.

13.5
13.07
13.2
13.09 Compare 13.07 and 13.09: $13.07 < 13.09$

Solution: B. Conor's time of 13.07 is the lowest, so he is the fastest.

13. Solution:

Values Less Than $\frac{1}{2}$, 0.5, or 0.50	Values Greater Than $\frac{1}{2}$, 0.5, or 0.50, and Less Than 1, 1.0, or 1.00	Values Greater Than 1, 1.0, or 1.00
$\frac{1}{4}$ (if $\frac{2}{4} = \frac{1}{2}$; $\frac{1}{4} < \frac{1}{2}$)	$\frac{7}{8}$ (if $\frac{4}{8} = \frac{1}{2}$; $\frac{7}{8} > \frac{1}{2}$)	$\frac{3}{2}$ (if $\frac{2}{2} = 1$; $\frac{3}{2} > 1$)
0.09 (0.09 < 0.50)	**0.6** (0.6 > 0.5)	$1\frac{1}{3}$ ($1\frac{1}{3} > 1$)
0.44 (0.44 < 0.50)		**1.01** (1.01 > 1.00)

Chapter Checklist for Standards

If you answered the questions correctly, you are on your way toward mastering the concepts and skills for this standard.

Standard	Concept/Skill	Questions
4.NF.1	I can recognize and generate equivalent fractions.	1, 6, 10
4.NF.2	I can compare two fractions with different numerators and different denominators by comparing to benchmark fractions.	9
4.NF.3	I can build fractions from unit fractions and add/subtract like fractions.	2, 3, 4, 7, 8
4.NF.4	I can apply previous understandings of multiplication to multiply a fraction by a whole number.	5
4.NF.5	I can express fractions using decimal notation.	13
4.NF.6	I can use decimal notation for fractions.	11
4.NF.7	I can compare decimal amounts.	12

Measurement and Data (MD)

CHAPTER 6

A measurement basically answers the question "How much?" We measure an object to find an amount of an object with questions such as "How big?", "How long?", and "How heavy?" Tools such as rulers, protractors, scales, and containers are used to determine a specific type of measurement. We use two different measurement systems to find the "How much?" of an object such as length, weight, mass, and volume. For nonsolid objects, such as air, we measure the volume or temperature. We also measure time to find either the current time or a length of time, usually in hours or minutes. The chart below shows the types of systems and units of measurement for each system. Note that the units are listed in order from smallest to largest unit.

Metric System	Measurement	Customary/English System
centimeter (cm) meter (m) kilometer (km)	length/distance	inch (in) foot (ft) yard (yd) mile (mi)
gram (gm) kilogram (kg)	mass/weight	ounce (oz) pound (lb) ton (T)
milliliter (mL) liter (L) (l)	volume/liquid capacity	fluid ounce (fl oz) cup (c) pint (pt) quart (qt) gallon (gal)

Length and Distance

We measure length (for shorter amounts) and distance (for longer amounts) in a straight line from one point to another, in any direction.

Smaller measurements are measured in centimeters and inches. They are used to measure things like line segments, pencils, paper clips, and small animals. Medium-sized measurements are measured in feet, yards, and meters. They are used to measure things like your height, the length of a room, a car, a field, or a larger animal. Larger measurements, such as miles and kilometers, are used to measure distances between towns and states, or the length of a river, or even the distance from the Earth to the Moon.

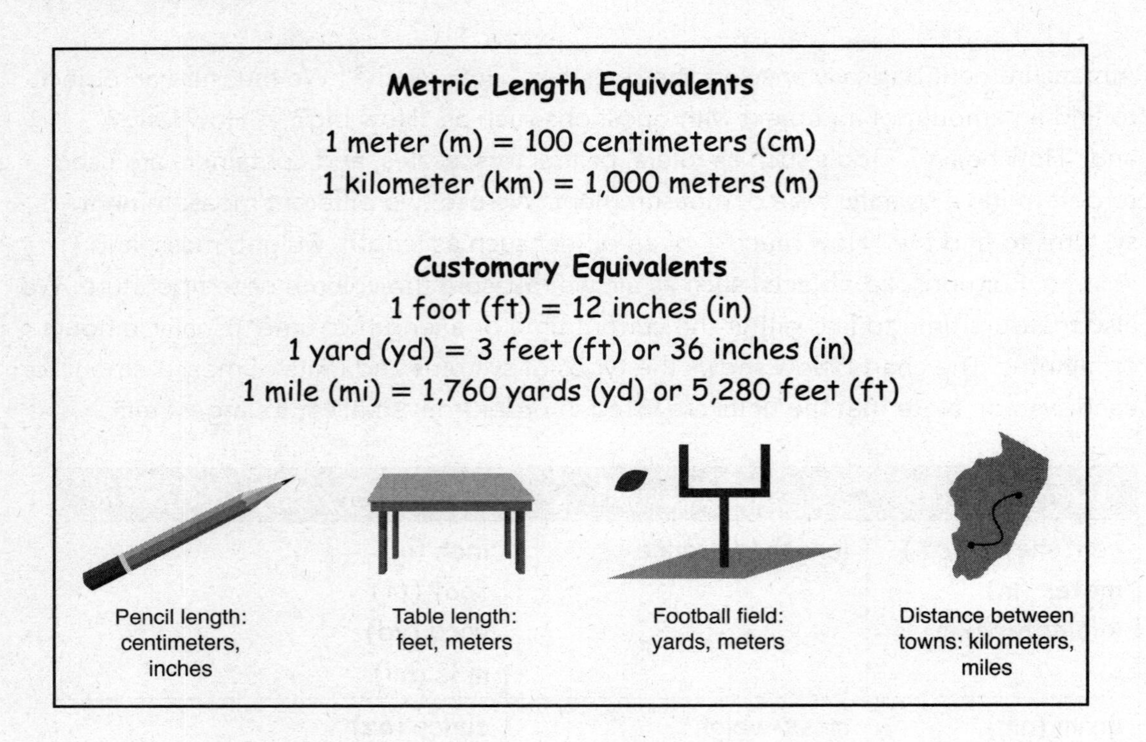

Finding Length Using a Ruler

1. Determine whether you are measuring centimeters (cm) or inches (in) by checking the label on the ruler.
2. Count the spaces between the labeled units to find the value of each space.
3. Be sure to evenly place the left edge of the object with the 0 on the ruler. On some rulers, the 0 is at the left edge, on others, it is not.
4. Compare the end of the object to the actual or closest marked measurement to find the actual or estimated measurement.

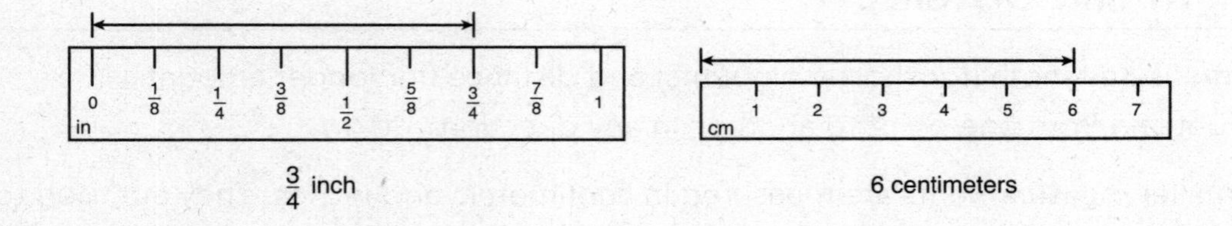

Mass and Weight

Mass is the amount of matter or space that an object takes up. We use the metric system to measure smaller objects with grams and larger objects with kilograms. Weight is a measure of how light or heavy an object is. We use the customary system to measure smaller objects with ounces. Medium-sized objects are measured with pounds, and larger objects are measured with tons.

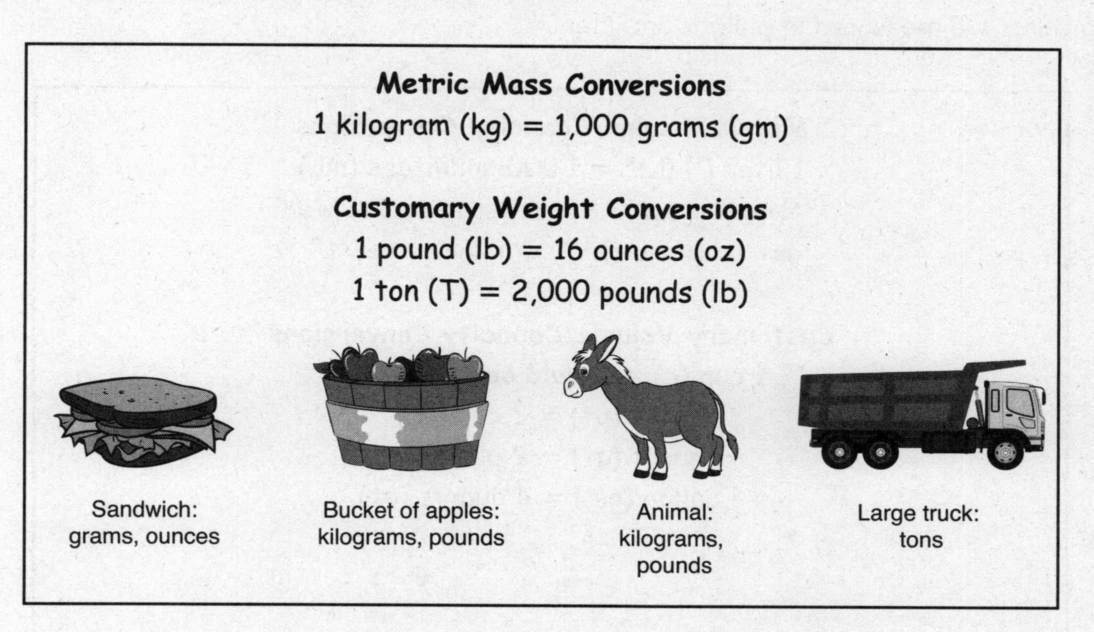

Finding Weight or Mass Using a Scale

1. Determine whether you are measuring ounces/pounds or grams/kilograms by checking the label on the scale.
2. Count the spaces between labeled units to find the value of each space.
3. Compare the dial to the actual or closest marked measurement to find the actual or estimated weight/mass.

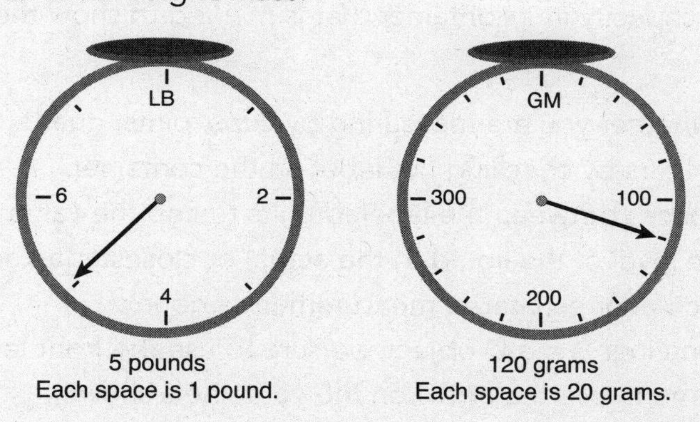

5 pounds
Each space is 1 pound.

120 grams
Each space is 20 grams.

Volume and Liquid Capacity

Volume is the amount of space something takes up. We use both systems of measurement to find the volume of 3-D objects or the amount of liquid in a container, which is also a 3-D object. Liquid capacity is the amount of liquid in a container, or how much the container can hold. Both volume and liquid capacity can be measured using smaller measures such as milliliters, fluid ounces, cups, pints, and quarts. Larger amounts are measured in gallons and liters.

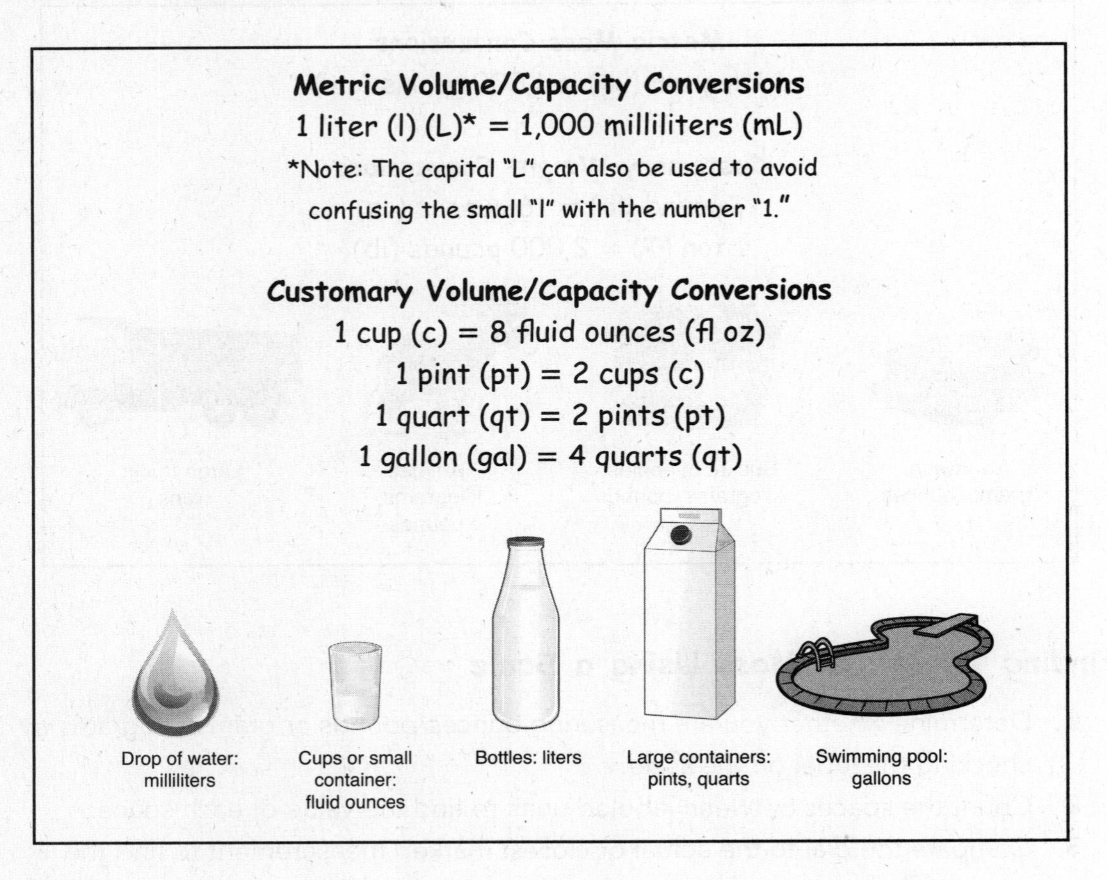

We measure liquid capacity in a container that is marked to show the unit of measurement.

1. Determine whether you are measuring by cups, pints, quarts, gallons, milliliters, or liters by checking the label on the container.
2. Count the spaces between the labeled units to find the value of each space.
3. Compare the level of the liquid to the actual or closest marked measurement to find the actual or estimated measurement amount.
4. Since the container is a 3-D object, be sure to use the front level line to find the measurement (see the arrow on the containers shown).

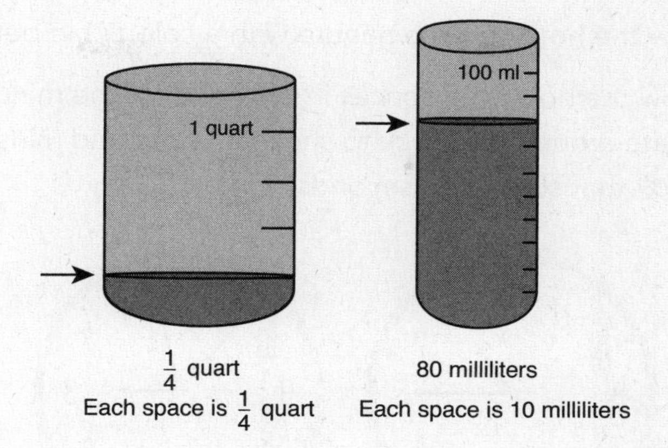

Time

We tell time by looking at the current reading on a clock. We measure time by finding the length of time from one point in time to another. Time is the same for both measurement systems. The units of time are the second, minute, and hour. Longer units of time are the day, month, and year. In this section, we use the second, minute, and hour units.

Time Equivalents
1 minute (min) = 60 seconds (sec)
1 hour (hr) = 60 minutes (min)

Telling Time

The current time is shown in hours and minutes. The hours are between 1 and 12 and the minutes are between 1 and 59. Between the hours of midnight (12:00) and 1 minute before noon (11:59), the time is known as A.M. Between the hours of noon (12:00) and 1 minute before midnight (11:59), the time is known as P.M. There are two kinds of clocks that are used to show the time:

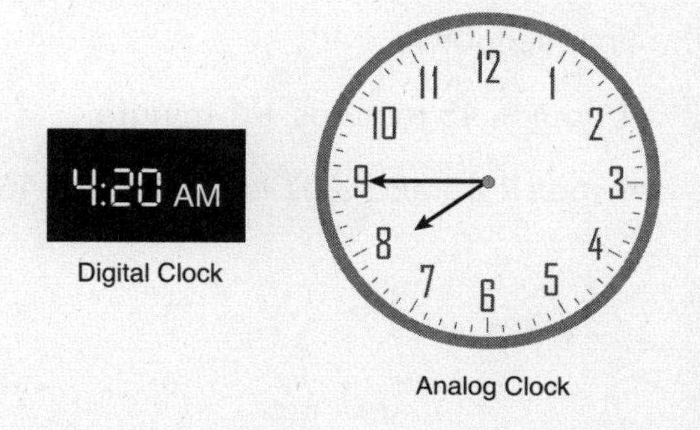

Digital clocks show the hour and the minute with a colon (:) in between.

Analog clocks show the hour with spaces in between for the minutes. On the clock are "hands" that rotate around the clock to show the hour and minute. Some clocks have a "second hand" that shows the seconds.

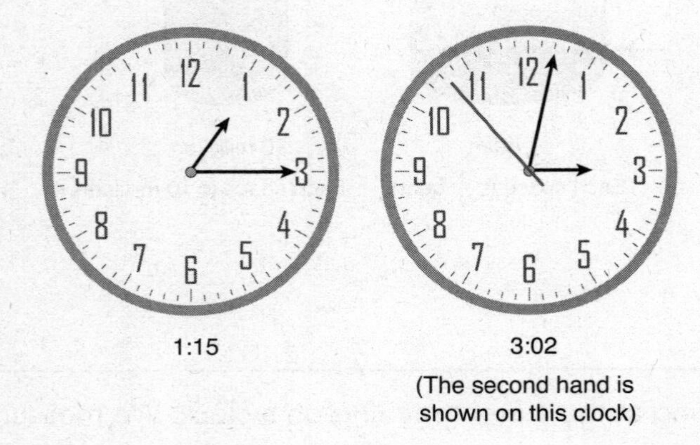

1:15

3:02
(The second hand is shown on this clock)

On the clocks above, the numbers are for the hours. The spaces between the hours are the minutes. You can quickly count the minutes by 5s when counting the hours. For example, for the time 1:15, the minute hand is on the 3, so you can count by 5s (5, 10, 15) to find the number of minutes after 1:00.

Measuring Time

We measure elapsed time by finding the difference between two times: the starting time and the ending time.

When measuring elapsed time that is less than 1 hour, use minutes.

For longer times, use hours and minutes, breaking the time into sections.

1. To find the elapsed time between 1:15 (starting time) and 3:01 (ending time), we first count the full hours. 1:15 to 2:15 is 1 full hour.
2. Count the minutes between 2:15 and the next hour. There is a total of 45 minutes between 2:15 and 3:00. Then, count the minutes between the next full hour and the ending time. There is 1 more minute.
3. Find the sum of the time sections:

   1 hour + 45 minutes + 1 minute

   The elapsed time between 1:15 and 3:01 is 1 hour and 46 minutes.

When writing elapsed time, avoid writing it to look like the current time. 1:46 minutes can be confused with the time 1:46.

In some cases, elapsed time may be displayed on a number line. The measurement appears just like a number line, measuring from left to right.

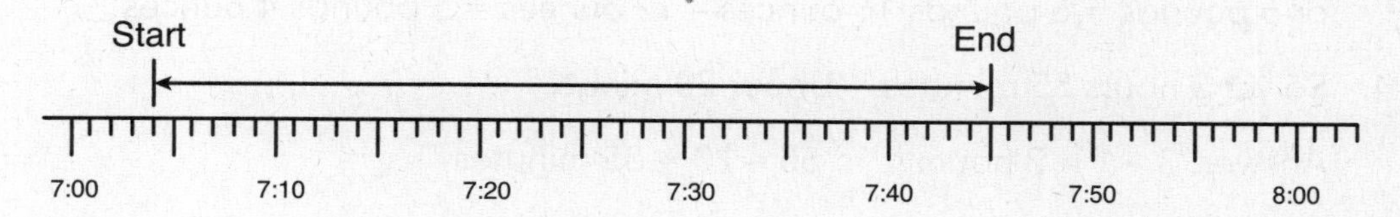

Shane played his drums from the start time to the end time shown on the number line. Shane played his drums from 7:04 to 7:45, for a total of 41 minutes. You can find the elapsed time by counting on the number line. Since the time does not go past the next hour, you can also subtract the minutes: 45 – 4 = 41. If the time went past the next hour, you would not be able to subtract the minutes.

Solving Word Problems Using Measurement

We can apply our understanding of units of measurement to solve problems involving the four operations of addition, subtraction, multiplication, and division. When the operations involve different units within the same system, measurements must be converted to similar units. Since the metric system is a base ten system, we can use our place value properties when solving problems involving operations. The customary system is not a base ten system, which will involve converting units of measurement when problem solving. Fractions and decimals can be used for the measurements.

Addition/Subtraction Examples

1. Solve: $3\frac{1}{2}$ meters + 20 centimeters = ? centimeters

   Answer: 3 m × 100 = 300 cm; $\frac{1}{2}$ meter = 50 cm; 350 + 20 = 370 cm

2. Solve: 4 feet + 5 inches = ? inches

   Answer: 4 ft × 12 = 48 in   48 + 5 = 53 inches

3. Solve: 6 pounds – 12 ounces = ? ounces or ? pounds ? ounces

   Answer: 6 pounds = 6 × 16 = 96 ounces   96 – 12 = 84 ounces

   or 6 pounds = 5 pounds 16 ounces – 12 ounces = 5 pounds 4 ounces

4. Solve: 3 hours 55 minutes – 1 hour 20 minutes

   Answer: 3 – 1 = 2 hours   55 – 20 = 35 minutes

   2 hours, 35 minutes

Multiplication/Division Examples

1. Solve: 3 gallons = _____ quarts

   Answer: 1 gallon = 4 quarts   3 × 4 = 12 quarts

2. Solve: How many 2-foot pieces of rope can be cut from a rope with a length of 4 yards? 1 yard = 3 feet

   Answer: 4 yards × 3 = 12 feet   12 ÷ 2 = 6 pieces of rope

3. Solve: 1.5 liters ÷ 3 = _____ milliliters

   Answer: 1 liter = 1,000 mL; 0.5 liters = 500 mL   1,500 mL ÷ 3 = 500 mL

Measurement Practice Problems

1. Measure the objects shown below. Write the numbers and words shown in the blanks to correctly complete the measurements. You can use the numbers and words more than once. Be sure to use the correct unit of measurement.

A.

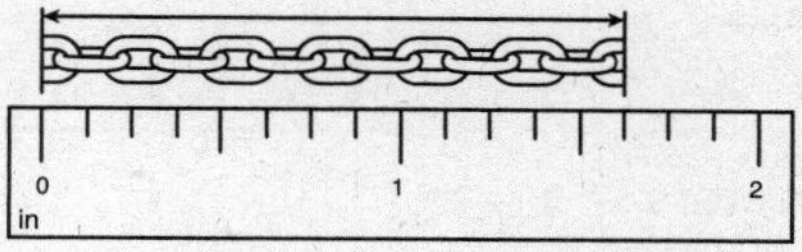

0 1 2
in

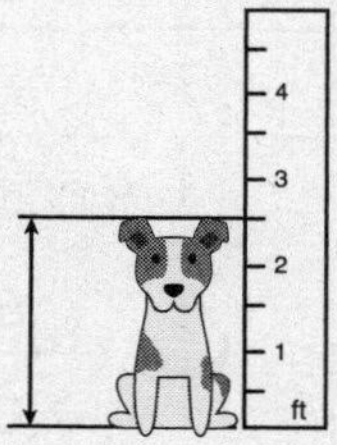

B.

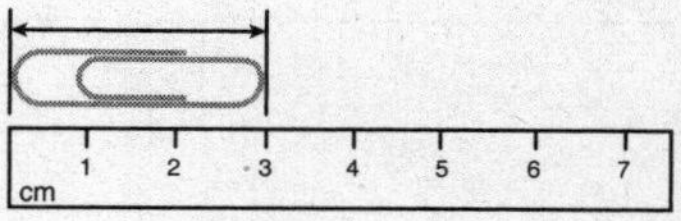

1 2 3 4 5 6 7
cm

C.

4
3
2
1
ft

D.

1 2 3 4 5 6 7
m

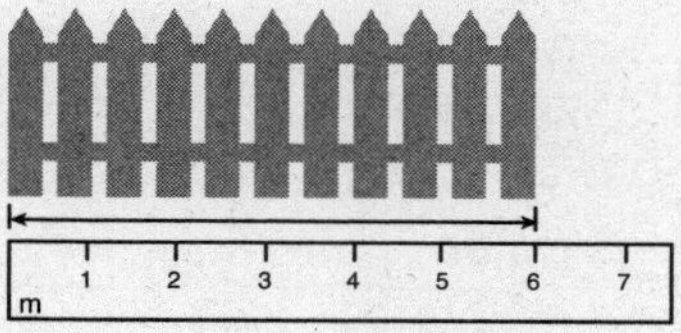

0 1 2 3 4 5 6 7 8 9

inches feet meters yards centimeters

A.

C.

B.

D.

2. The chart below shows some measurements and their equivalents. Circle T if the equivalent is true and F if the equivalent is false.

First Measurement	Equivalent Measurement	True (T) or False (F)
A. 2 feet	24 inches	T F
B. 4 yards	16 feet	T F
C. 5 meters	5,000 centimeters	T F
D. 5 feet	60 inches	T F
E. 7 kilometers	7,000 meters	T F
F. 1 foot 3 inches	15 inches	T F
G. $3\frac{1}{2}$ meters	350 centimeters	T F
H. 2 miles	3,520 feet	T F

3. Complete the sentences by circling the unit of measurement that is the most reasonable.

A. A table has a height of about 3 ______.

inches feet centimeters

B. A piece of paper has a length of about 11 ______.

inches yards meters

C. The distance from Atlantic City to Jersey City is about 120 ______.

meters yards miles

D. A cell phone has a length of about 12 ______.

feet miles centimeters

4. Kaitlyn has a new puppy. She measured the puppy and said that its height was 12. What is a reasonable unit of measurement that could be the height of Kaitlyn's puppy?

5. Place the object names and amounts in the table to show the most reasonable unit of measurement that would be used to measure the mass or weight.

10 paper clips	10 bananas	10 pencils	10 cars	100 books	100 elephants

Grams or Ounces	Pounds or Kilograms	Tons

6. The scale shows the weight and mass of two different objects. Find the correct measurement by circling the letter. Choose all that are correct.

Scale A

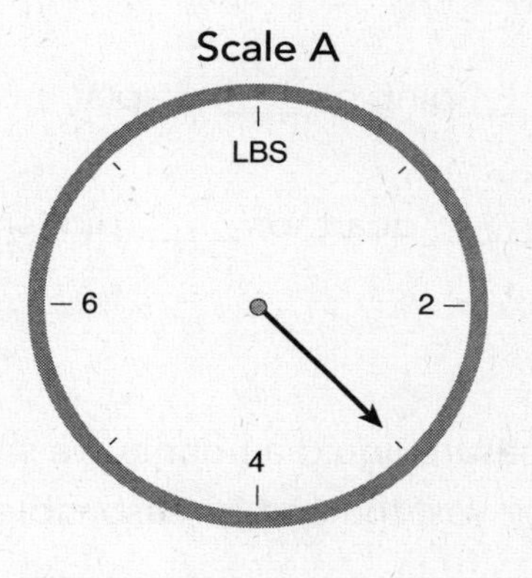

Scale B

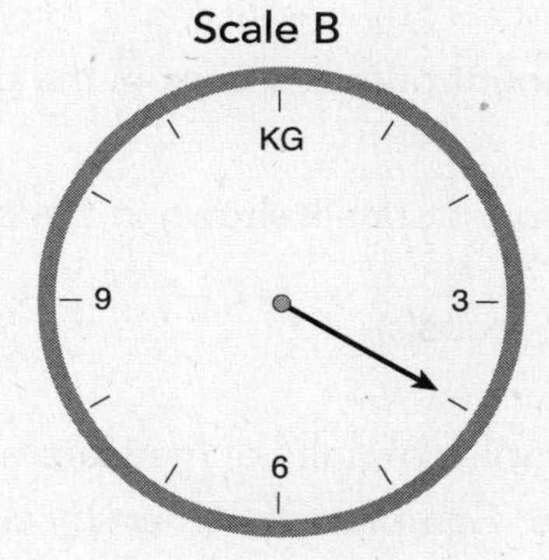

Scale A:

☐ A. 2.5 pounds
☐ B. 3 pounds
☐ C. 30 pounds
☐ D. 48 ounces
☐ E. 250 ounces
☐ F. 480 ounces

Scale B:

☐ A. 3.5 kilograms
☐ B. 4 kilograms
☐ C. 31 kilograms
☐ D. 300 grams
☐ E. 400 grams
☐ F. 4,000 grams

7. Jason placed some books and school supplies in his backpack. While walking to school carrying his backpack, he guessed that it was so heavy that it must have weighed about 20 ounces. Is Jason's guess reasonable? Why or why not?

8. Ted poured some milk from a pitcher into a glass as shown:

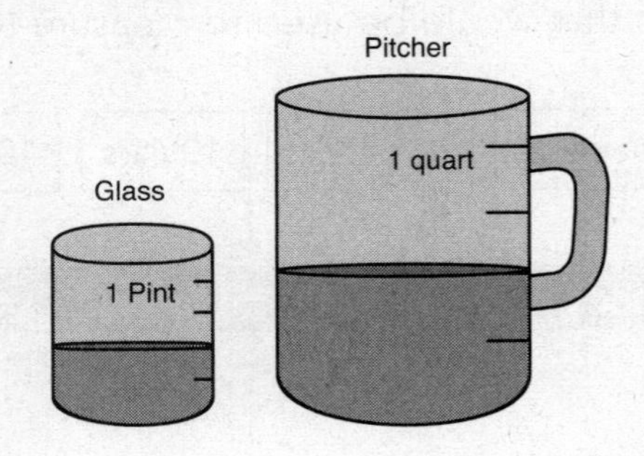

Write the numbers shown in the blanks to correctly complete each statement. You may use some numbers more than once or not at all.

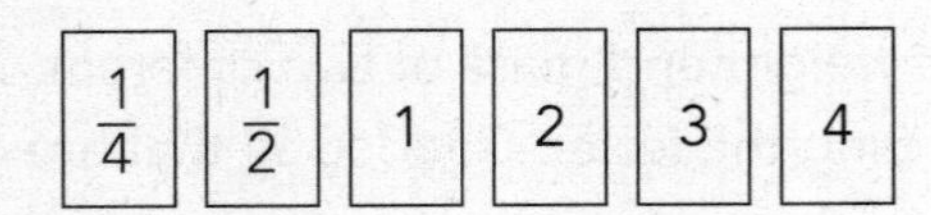

A. The amount of milk shown in the glass is _____ pint, or _____ cup(s).

B. The amount of milk shown in the pitcher is _____ quart, or _____ pint(s)

or _____ cups(s).

9. Some examples of units of measurement to measure liquid amounts are shown in the table. Complete the chart by circling Y for yes, the unit is reasonable, or N for no, the unit is not reasonable.

Example	Unit of Measurement	Reasonable (Y) or (N)	
A. Soup in a small bowl	Quarts	Y	N
B. Water in a large fishbowl	Liters	Y	N
C. Water in a pool	Pints	Y	N
D. Small bottle of juice	Milliliters	Y	N
E. Water in a lake	Gallons	Y	N

10. Nazia has a 2-liter bottle of juice. The bottle is a little more than half full. What could be a reasonable amount of juice, in milliliters, that Nazia has in the bottle? Explain how you found your answer.

11. The table shows how long it took some students to complete their math homework. The students and their starting and ending times are shown:

Student	Starting Time	Ending Time
Jamal	4:48	5:15
Derek	4:06	4:41
Mia	4:57	5:25
Erica	5:03	5:39

Place the student names in order from the least to greatest length of time that it took them to complete their math homework.

__________ __________ __________ __________

least greatest

12. The length of Allison's math class is 53 minutes. The start time is shown on the number line below. What time does Allison's math class end? Place an X on the number line to show the time that Allison's math class ends. Write the start and end times on the blanks below.

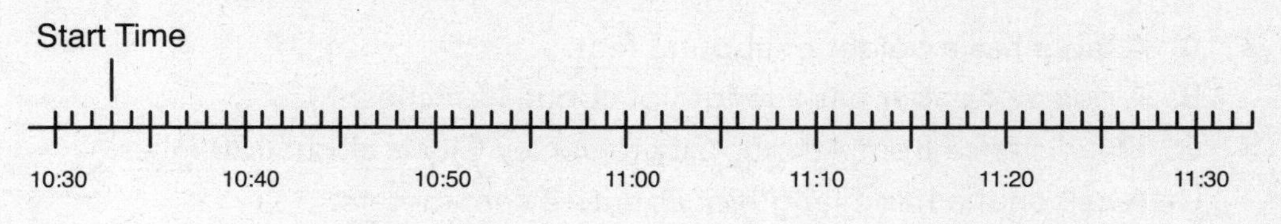

Start time: ________ End time: ________

13. Alexa has 2 pieces of ribbon. They measure $2\frac{1}{2}$ feet and 3 feet. How many total inches of ribbon does Alexa have? Choose the correct answer.

O A. 65 inches
O B. 66 inches
O C. 85 inches
O D. 90 inches

14. Elaria's mom is making milk shakes. She is filling 2 glasses each with an amount of 0.7 liters. What is the total amount of milliliters in the 2 glasses combined?

____

Answers to Measurement Practice Problems

1. A. $1\frac{5}{8}$ inches. There are 8 spaces between 1 and 2 inches. The measurement ends at the fifth space. The ruler is labeled "in."
   B. 3 centimeters. The ruler is labeled "cm."
   C. $2\frac{1}{2}$ feet. The measurement ends right in the middle of 2 and 3. The ruler is labeled "ft."
   D. 6 meters. The ruler is labeled "m."

2. A. T: $2 \times 12 = 24$
   B. F: 4 yards equals 12 feet. $4 \times 3 = 12$
   C. F: 5 meters = 500 centimeters. $5 \times 100 = 500$
   D. T: 5 feet = 60 inches. $5 \times 12 = 60$
   E. T: 7 kilometers = 7,000 meters. $7 \times 1{,}000 = 7{,}000$
   F. T: 1 foot = 12 inches. 12 + 3 inches = 15 inches
   G. T: 3 meters = 300 centimeters. $\frac{1}{2}$ meter = 50 centimeters
   $300 + 50 = 350$
   H. F: 2 miles = 10,560 feet. $2 \times 5{,}280 = 10{,}560$ (2 miles = 3,520 yards not feet)

3. A. A table has a height of about 3 feet.
   B. A peice of paper has a length of about 11 inches.
   C. The distance from Atlantic City to Jersey City is about 120 miles.
   D. A cell phone has a length of about 12 centimeters.

4. The most reasonable unit would be 12 inches for a puppy.

5. Grams/ounces: 10 paper clips, 10 pencils
   Pounds/kilograms: 10 bananas, 100 books
   Tons: 10 cars, 100 elephants

6. Scale A: **B** 3 pounds and **D** 48 ounces
   Scale B: **B** 4 kilograms and **F** 4,000 grams

7. No. Jason's guess is not reasonable. A very heavy backpack would feel more like 20 pounds or 20 kilograms. A weight of 20 ounces would not feel very heavy.

8. A. The amount of milk shown in the glass is $\frac{1}{2}$ pint, or 1 cup.
   B. The amount of milk shown in the pitcher is $\frac{1}{2}$ quart, or 1 pint or 2 cups.

9. A. Soup bowl in quarts: N
   B. Fishbowl in liters: Y
   C. Pool in pints: N
   D. Small bottle in milliliters: Y
   E. Lake in gallons: Y

10. The most reasonable measure of milliliters for a little more than half of a 2-liter bottle would be a measurement between 1,001 mL and 1,250 mL. A measurement between 1,251 and 1,499 is a possible answer, but not the most reasonable.

11. Jamal (27 min), Mia (28 min), Derek (35 min), Erica (36 min)

12. The X should be placed on 11:26.
    Start Time: 10:33
    End Time: 11:26
    This can be solved a few ways:

    Strategy 1: There are 27 minutes between 10:33 and 11:00.

    $$53 - 27 = 26$$

    The rest of the time is from 11:00 to 11:26, so the class ends at 11:26.

    Strategy 2: 60 minutes – 53 minutes = 7 minutes.
    One hour from 10:33 is 11:33.

    $$33 - 7 = 26$$

    So the class ends at 11:26.
    Strategy 3: Count on the number line.

13. **B** 66 inches
    2 ft = 24 in
    $\frac{1}{2}$ ft = 6 in
    3 ft = 36 in
    24 + 6 + 36 = 66

14. 1,400 milliliters total
    0.7 + 0.7 = 1.4 liters
    1 liter = 1,000 mL
    0.4 liter = 400 mL
    1,000 + 400 = 1,400

Measurement of Area and Perimeter

To review, we can find the area of a 2-D or flat shape by measuring the amount of space that the shape covers. This measurement is in the form of square units, since the shape has 2 dimensions: a length and a width. We can find the perimeter of a 2-D or flat shape by measuring the sides around the outside of the shape. This measurement is in the form of a unit of length, since the distance around is basically a line segment.

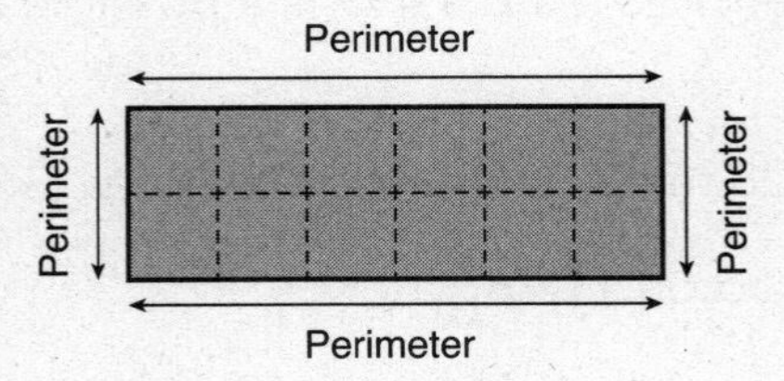

We see that the shape is covered with 12 unit squares giving the shape an area of 12 square units or 12 units squared. We see that the side length across the top of the shape is 6 units and the side width is 2 units. We add 6 + 2 + 6 + 2 to find the perimeter of 16 units. We use the term *units* when the unit of measurement is not known. We also see that the squares completely cover the shape without gaps or overlaps.

Solving for Area and Perimeter by Using Formulas

When the unit squares are not shown, we can find the area and perimeter by using formulas.

$$\text{Area} = \text{length} \times \text{width}$$

8 cm

4 cm

$$A = 8 \times 4 = 32 \text{ sq cm}$$

or

$$\text{Perimeter} = \text{length} + \text{width} + \text{length} + \text{width}$$

$$P = 8 + 4 + 8 + 4 = 24 \text{ cm}$$

Additional Perimeter Formula

$$(2 \times L) + (2 \times W) = 16 + 8 = 24 \text{ cm}$$

Solving for Unknown Sides

When the area or perimeter of a shape is known, we can work backward to find the length of the unknown sides.

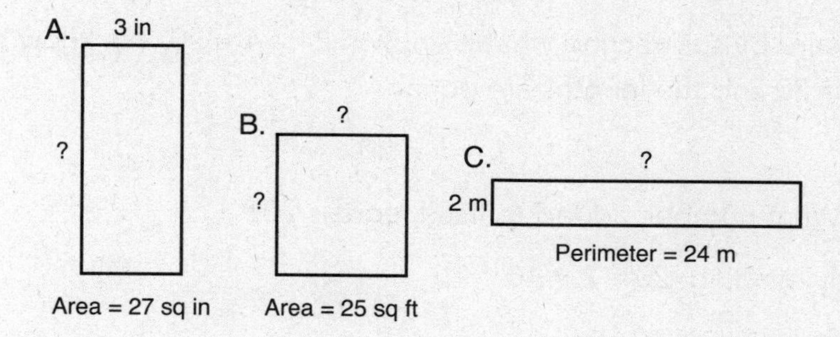

Example A

We can start by using the formula and placing what is known in the formula:

$$A = L \times W$$
$$27 = 3 \times ?$$

We can divide to find the unknown width:

$$27 \div 3 = 9$$

The width is 9 inches. We can multiply to check:

$$3 \times 9 = 27$$

Example B

If the shape is a square, we know that the length and width are equal:

$$L \times W = 25$$
$$25 \div ? = ?$$

We can use guess and check to see which number, when multiplied by itself, equals 25:

$$2 \times 2 = 4$$
$$3 \times 3 = 9$$
$$4 \times 4 = 16$$
$$5 \times 5 = 25$$

We see that $5 \times 5 = 25$, so the length and the width are both 5 feet.

Example C

We can start with the formula $L + 2 + L + 2 = 24$.

$$4 + L + L = 24$$

We work backward by subtracting what is known: $24 - 4 = 20$. We know that the sum of the lengths is 20 and the lengths are equal:

$$L + L = 20$$

Ask yourself: "What number, added to itself, equals 20?"

We can solve by dividing: $20 \div 2 = 10$

The missing side length is 10 meters.

Check by using the formula $10 + 2 + 10 + 2 = 24$.

We can find the area of nonrectangular shapes by cutting the shapes into rectangles, finding the area of each shape, and adding to find the total area.

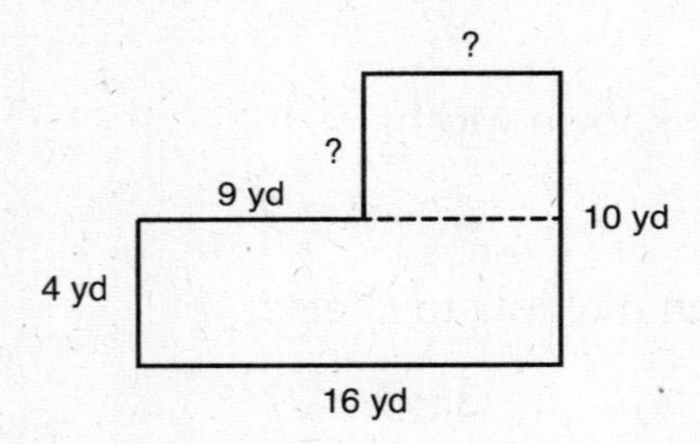

The larger rectangle: $L = 16$ yd and $W = 4$ yd

Area = 64 sq yd

The smaller rectangle: $L = (10 - 4) = 6$ yd and $W = (16 - 9) = 7$ yd

Area = $6 \times 7 = 42$ sq yd

Total area = $64 + 42 = 106$ sq yd

Real-World Problem Solving with Area and Perimeter

Area and perimeter are used in many real-world situations.

Examples

Objects to measure: table, mat, floor, wall, backyard, picture frame, bulletin board, roof

Area problems include measuring to find an area to paint, to place square tiles, paper, grass, or carpet. Measurements are of the inside space.

Perimeter problems include measuring the perimeter to place a fence, or a border or frame around something. Measurements are taken around the outside of the space.

Measurement of Area and Perimeter Practice Problems

1. A square has a side length of 8 inches. What is the area and the perimeter of the square? Place the numbers and words in the spaces provided to correctly complete the statements below:

16	32	64	inches	square inches

The area of the square is ________ ________.

The perimeter of the square is ________ ________.

2. Both shapes have an area of 36 square inches. Which THREE measurements could be the length and width of the shapes?

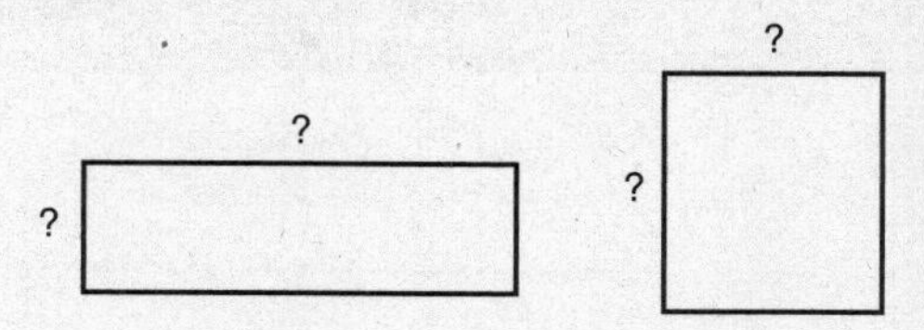

☐ A. 3 inches by 12 inches
☐ B. 3 inches by 15 inches
☐ C. 4 inches by 9 inches
☐ D. 4 inches by 14 inches
☐ E. 6 inches by 6 inches
☐ F. 6 inches by 12 inches

3. Mr. Ramos is placing some square foot tiles on the floor. He has a total of 135 square tiles. The length of the floor is 9 feet.

PART A

What is the width of the floor?

PART B

What is the perimeter of the floor? Explain how you found your answer.

4. The table shows one perimeter and one area measurement. Complete the table by placing the length and width that can be used to find the perimeter or area. Some may be used more than once or not at all.

$L = 2$; $W = 16$	$L = 3$; $W = 16$	$L = 4$; $W = 12$
$L = 6$; $W = 8$	$L = 8$; $W = 8$	$L = 8$; $W = 16$

Perimeter = 32 Units	Area = 48 Square Units

5. Kate's closet floor has a length of 3 feet and a width of $1\frac{1}{2}$ feet.

Complete the statement by placing your answer on the line below.

The perimeter of the floor in Kate's closet is __________.

6. A triangle has equal sides. Each is 0.6 meter long. Which shows the perimeter of the triangle?

O A. 0.18 m
O B. 0.24 m
O C. 1.8 m
O D. 2.4 m

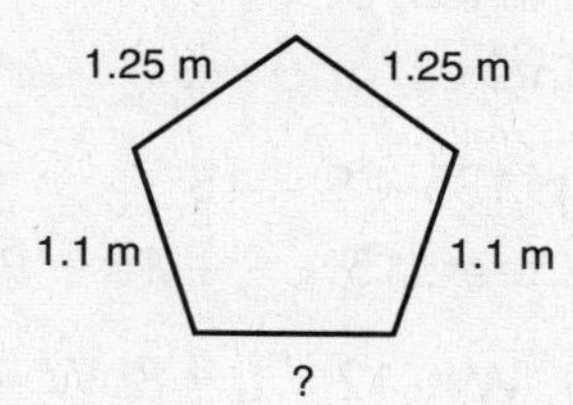

7. The perimeter of the shape is 5 meters. Renee said that the missing side measurement is 0.3 meters. Is Renee correct? Why or why not?

---

8. The length and width of a bulletin board is 18 feet by 4 feet. Michelle wants to place square feet posters on the board. She has a total of 70 posters. Will Michelle be able to completely cover the board with the posters without gaps or overlaps? Explain why or why not.

Answers to Measurement of Area and Perimeter Practice Problems

1. Area = 64 square inches
   The sides are all 8 inches: $8 \times 8 = 64$
   Perimeter = 32 inches: $8 + 8 + 8 + 8 = 32$ inches

2. **A** Correct: 3 inches by 12 inches, or 3×12
   C Correct: 4 inches by 9 inches, or 4×9
   E Correct: 6 inches by 6 inches, or 6×6

3. Part A: The width is 15 feet: $135 \div 9 = 15$
   Part B: The perimeter is 48 feet: $15 + 9 + 15 + 9 = 48$

4. Perimeter = 32 units: $L = 4$; $W = 12$ $L = 8$; $W = 8$
   Area = 48 square units: $L = 3$; $W = 16$ $L = 4$; $W = 12$ $L = 6$; $W = 8$

5. The perimeter is 9 feet: $3 + 1\frac{1}{2} + 3 + 1\frac{1}{2} = 9$

6. **C** 1.8 m; $0.6 + 0.6 + 0.6 = 1.8$

7. Yes, Renee is correct. The sum of the 4 sides is 4.7.

   $$5 - 4.7 = 0.3 \text{ m}$$

8. No. Michelle will not be able to cover the board with 70 square feet posters:

   $$18 \times 4 = 72 \text{ sq ft}$$

   The area is 72 sq ft, which is more than 70.

Data: Representing and Interpreting Data with a Line Plot

Data sets provide mathematical information that can be displayed using graphs and tables. Many data sets include measurements that are displayed on graphs such as bar graphs, line graphs, picture graphs, and line plots. A line plot can be used to show a data set using a number line.

In grade 3, line plots displayed whole numbers. In grade 4, line plots can also display fractions. We can use line plots to find information about the data.

Example

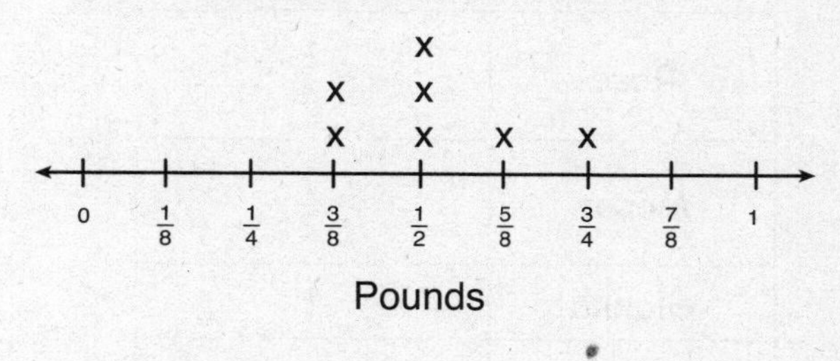

The top title of the line plot tells us that the weights of some bags of tomatoes were measured. The bottom title tells us that the unit of measurement used was pounds. The number line shows that there are 8 spaces between the 0 and 1, so each space counts up by $\frac{1}{8}$ pound.

We can use the line plot to determine the following information.

- A total of 7 weights are shown.
- The greatest weight is $\frac{3}{4}$ lb or $\frac{6}{8}$ lb.
- The least weight is $\frac{3}{8}$ lb.
- The most common weight is $\frac{1}{2}$ lb or $\frac{4}{8}$ lb.
- The difference between the greatest and least weight can be found by solving $\frac{6}{8} - \frac{3}{8} = \frac{3}{8}$ lb.
- The total weights of the bags can be found by solving

$$\left(2\times\frac{3}{8}\right)+\left(3\times\frac{4}{8}\right)+\frac{5}{8}+\frac{6}{8}=\frac{6}{8}+\frac{12}{8}+\frac{5}{8}+\frac{6}{8}=\frac{29}{8}=3\frac{5}{8}\text{ lb.}$$

Data Line Plot Practice Problems

1. Some fourth-grade students measured their heights and compared them with their height in third grade. The table below shows the height increases, in inches, for 7 students.

Student	Height Increase (Inches)
Jabari	$\frac{1}{2}$
Zack	$\frac{3}{4}$
Rose	$\frac{1}{4}$
Mason	$\frac{1}{2}$
Gianna	1
Kaydee	$1\frac{1}{4}$
Carlos	$\frac{3}{4}$

PART A

Construct a line plot using the data from the chart. Be sure to include the titles and label the number line.

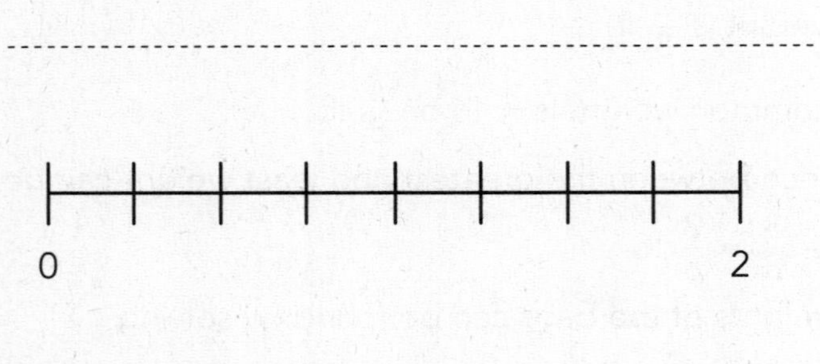

PART B

What is the difference, in inches, between the greatest and least height increase? Choose the correct answer.

O A. $\frac{1}{2}$ in

O B. 1 in

O C. $1\frac{1}{4}$ in

O D. $1\frac{1}{2}$ in

PART C

Another student, Liam, also measured his height increase. When Liam's height increase is included, the total height increase of the 8 students is $5\frac{3}{4}$ inches. What is Liam's height increase in inches? Explain how you found your answer.

Answers to Data Line Plot Practice Problems

1. Part A: Line plot construction

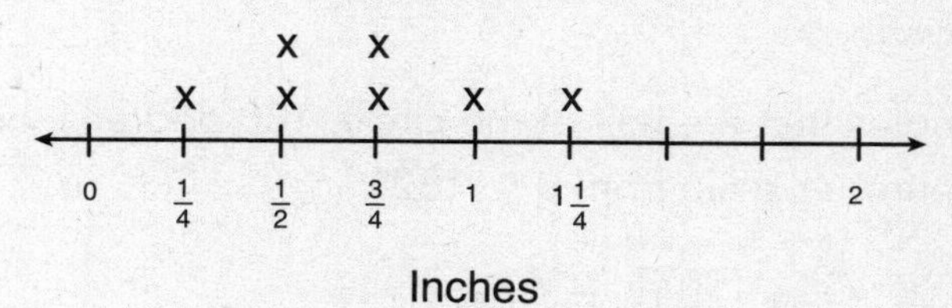

Part B: **B** Greatest height – least height: $1\frac{1}{4} - \frac{1}{4} = 1$

Part C: Liam's height increase is $\frac{3}{4}$ inch.

The sum of the 7 height increases = 5 inches. Subtract the total height increases of 8 students from the total height increases of 7 students:

$5\frac{3}{4} - 5 = \frac{3}{4}$

Geometric Measurement: Angles

Angle Review

An angle is formed by two rays with a common endpoint called a vertex.

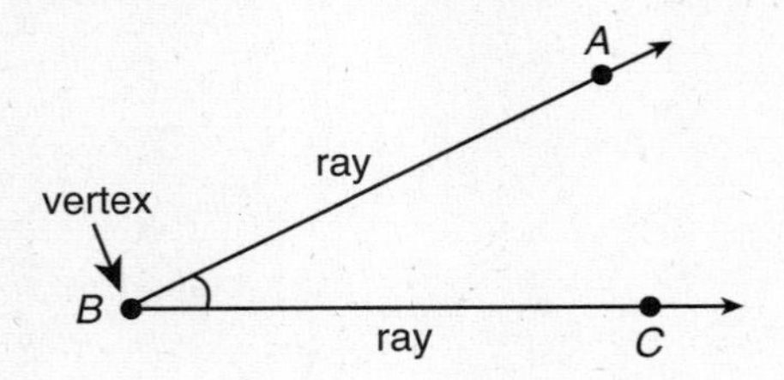

We can identify this angle as angle *ABC* or angle *B*.

We measure the angle as the opening between ray *BC* and ray *BA*. This is based on a circle as shown:

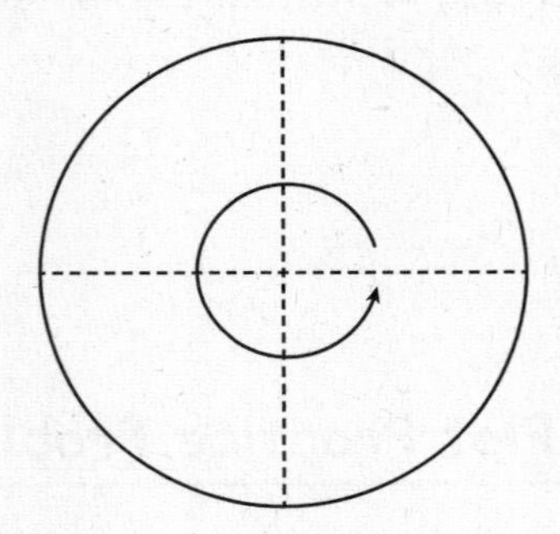

The circle consists of exactly 360, 1-degree measures. A full turn around the circle measures 360 degrees. Angles are measured by the number of 1-degree turns that the rays form on the circle.

We can measure angles that are half of the circle, 180 degrees, or less by using a protractor that has a number scale from 0 to 180°.

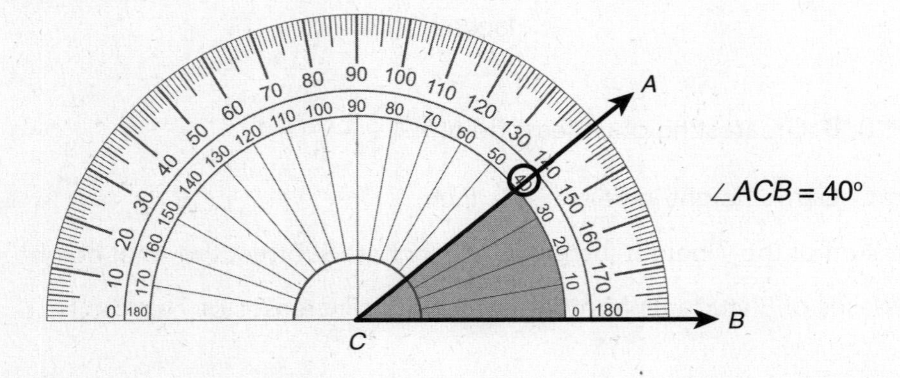

The protractor is used to measure angle *ACB* as 40 degrees, or 40°.

Finding Angle Measures

1. Place the protractor on the angle. Place the bottom center crossbar notch on the vertex (locate point *C* on the diagram).
2. Place one ray horizontally lining it up with the 0-degree measurement on the protractor.
3. Find the measurement by following the number scale from the 0 up to the measurement where the other ray crosses the number scale. For angles facing the right, the 0 is the bottom number scale. For angles facing the left, the 0 is the top number scale.

For angles larger than 180°, a full-circle protractor can be used.

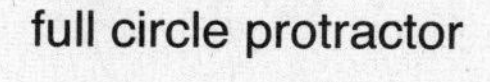

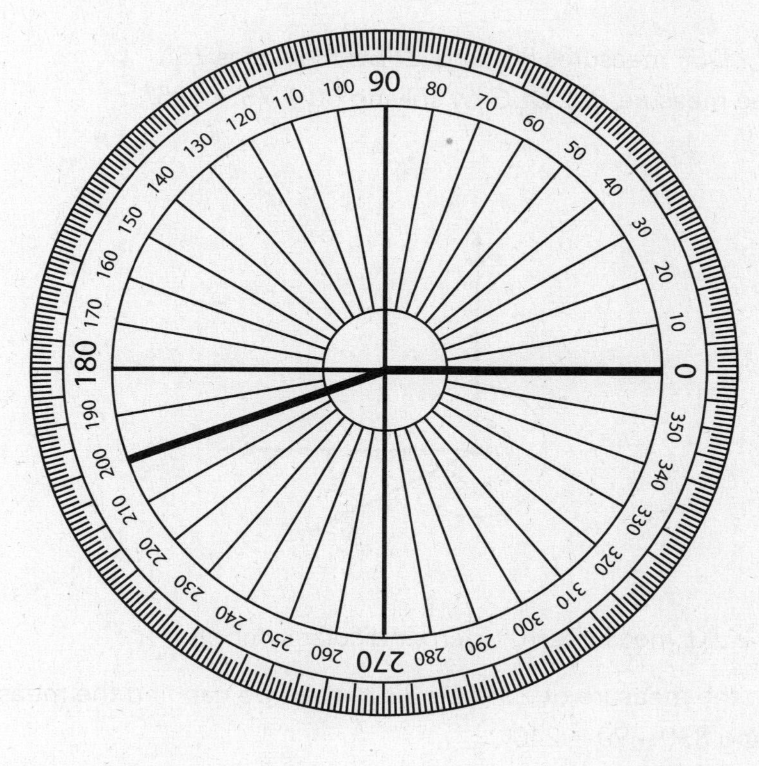

The angle shown has a measurement of 200°.

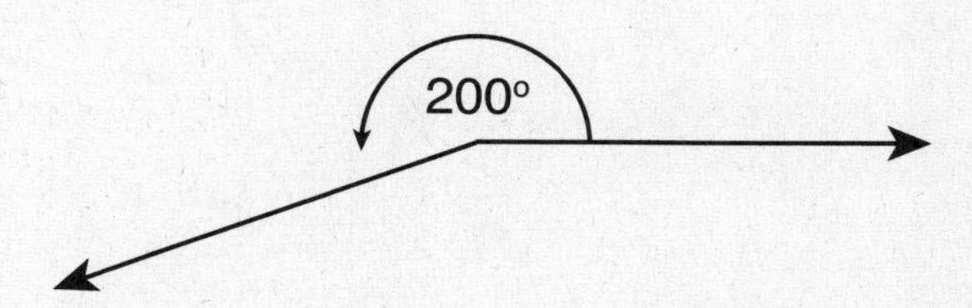

Using Addition and Subtraction to Find Angle Measures

Angles can be combined to form larger angles. We can find the unknown angle measures by adding to or subtracting from known angle measures.

Example A

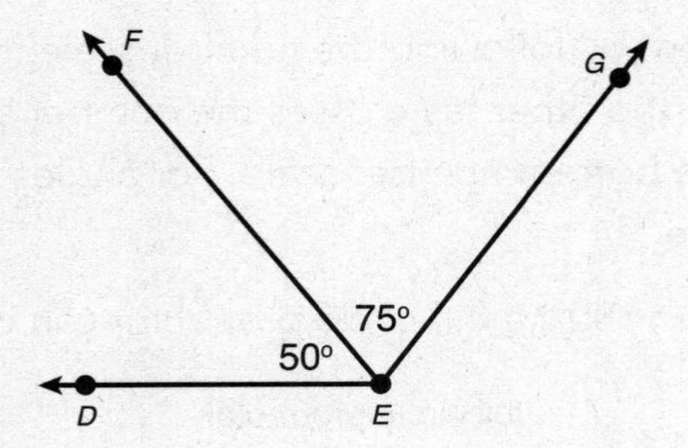

We know that $\angle DEF$ measures 50° and $\angle FEG$ measures 75°.
We can find the measure of $\angle DEG$ by solving $50 + 75 = 125°$.

Example B

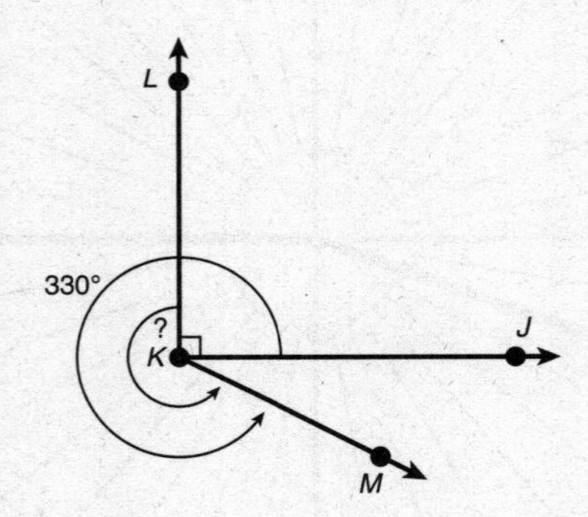

We know that $\angle JKL$ measures 90° (perpendicular symbol).

If we state that the measure of $\angle JKM$ is 330°, then we can find the measure of $\angle LKM$ by solving $330 - 90 = 240°$.

Angle Measurement Practice Problems

1. Place the correct angle measurements in the blanks. Some will not be used.

10°	40°	45°	55°	135°	140°	145°	170°

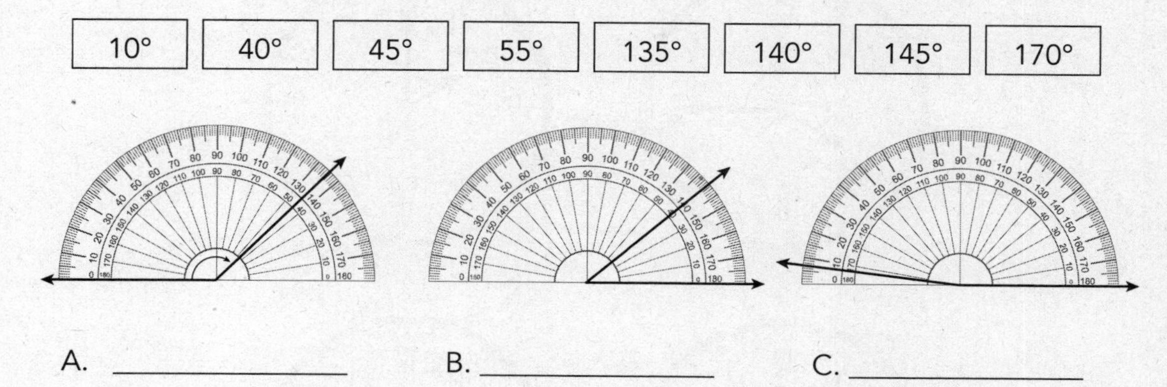

A. ______________ B. ______________ C. ______________

2. Use the diagram below to complete the table:

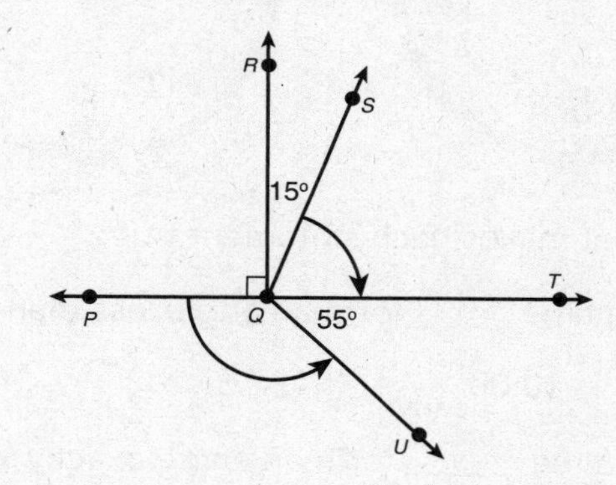

In the diagram, *PT* is a straight line. Find the missing angle measurements. Circle Y (yes) if the statement is correct and N (no) if the statement is not correct.

Statement	Correct: Yes (Y) or No (N)
A. ∠*PQS* measures 75°	Y N
B. ∠*SQT* measures 75°	Y N
C. ∠*PQS* measures 105°	Y N
D. ∠*SQT* measures 105°	Y N
E. ∠*PQU* measures 125°	Y N
F. ∠*RQU* measures 145°	Y N

3. The hands on the clocks below form angles.

Circle the answer that makes each statement true.

A. The hands of the time _____ form an angle less than 90°.

8:15 10:00 12:15

B. The hands of the time _____ form an angle exactly 90°.

8:15 10:00 12:15

C. The hands of the time _____ form an angle between 180° and 270°.

8:15 10:00 12:15

D. Draw a time on the last clock that forms an angle between 90° and 180°.

Answers to Angle Measurement Practice Problems

1. A. 135° B. 40° C. 170°
2. A. N B. Y C. Y D. N E. Y F. Y
3. A. 10:00 B. 12:15 C. 8:15
   D. Possible answers: 12:20–12:40; 1:25–1:45; 2:30–2:50; 3:35–3:55; 4:40–4:00/5:00; 5:45–5:05/6:05; 6:50–6:10/7:10; 8:00–8:20; 9:05–9:25; 10:10–10:30; 11:15–11:35

Measurement and Data Review Practice Problems

1. Which THREE measurements are equal to 3 feet?

☐ A. 1 yard
☐ B. 24 inches
☐ C. 2 feet 10 inches
☐ D. 30 inches
☐ E. 36 inches
☐ F. 2 feet 12 inches

2. Use the numbers shown to fill in the blanks to correctly complete each equation. You may use some numbers more than once or not at all.

2	4	8	10	12	16	24	32	48	100	1,000

2 ft = ______ in   2 lb = ______ oz

1 qt = ______ pt   1 gal = ______ qt

1 gal = ______ pt   1 kg = ______ g

1 m = ______ cm   1 L = ______ mL

3. Paige has ribbons of different colors and lengths as shown in the table below:

Ribbon Color	Ribbon Length
Red	2 feet 4 inches
Blue	20 inches
Pink	1 yard 2 inches
Green	30 inches
Yellow	2 feet 15 inches
Purple	2 feet 10 inches

Place the ribbon colors in order from the least to greatest length.

Red	Blue	Pink	Green	Yellow	Purple

______ ______ ______ ______ ______ ______

least greatest

Extended Performance Task

4. The Picnic: The Choi family of 8 had a picnic at the county park. Answer the following questions about the picnic.

PART A

Mrs. Choi brought enough fruit salad so that each of the 8 family members could eat exactly $\frac{1}{2}$ pound. Write an equation to show how to find the total amount of fruit salad. Solve your equation.

PART B

Mr. Choi brought 2 containers of iced tea for the family to drink. One container had a total of 3 liters. The other had 1,600 milliliters. After the picnic, there was 0.5 liters left. What is the total amount of iced tea, in milliliters, that the family drank during the picnic?

PART C

Mrs. Choi brought a tablecloth to place on a picnic table. The diagram below shows the area of the top of the picnic table.

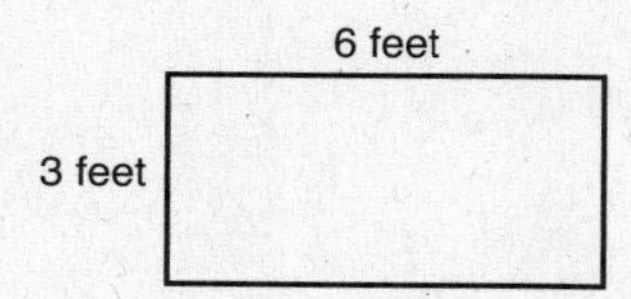

The length of the table cloth is 2 feet longer and 2 feet wider than the length and width of the picnic table. Write an equation to find the area of the tablecloth. Solve your equation.

PART D

The Choi family left their house at 9:15 A.M. It took them 25 minutes to get to the park. What time did they arrive at the park?

PART E

Mr. Choi brought his lounge chair to sit in and relax. The diagram below shows the two different chair positions, in an upright and in a reclining position.

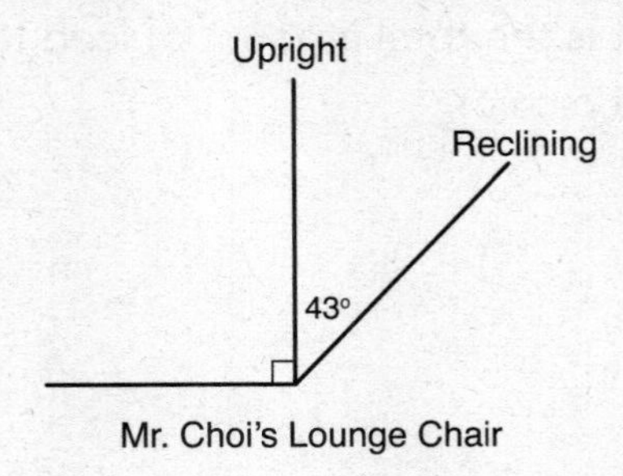

Mr. Choi's Lounge Chair

What is the total angle measurement of the chair in the reclining position? Explain how you found your answer.

5. Stella is learning how to juggle 3 balls. While she practices, Stella records how long she can juggle without dropping a ball. The line plot shows the last 10 juggling times.

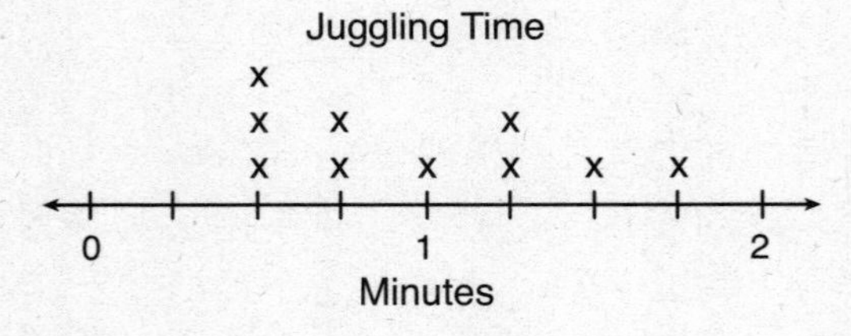

Circle T (true) or F (false) to correctly complete each statement shown.

A. The difference between the greatest and least time is $1\frac{1}{4}$ minutes. T F

B. The total of the times less than 1 minute is 4 minutes. T F

C. The total of the times more than 1 minute is 5 minutes and 45 seconds. T F

Answers to Measurement and Data Review Practice Problems

1. **A** Correct: 1 yard
   E Correct: 36 inches
   F Correct: 2 feet 12 inches

2. 2 ft = 24 in
   1 qt = 2 pt
   1 gal = 8 pt
   1 m = 100 cm
   2 lb = 32 oz
   1 gal = 4 qt
   1 kg = 1,000 g
   1 L = 1,000 mL

3. Ribbons in order: blue (20 in), red (28 in), green (30 in), purple (34 in), pink (38 in), yellow (39 in)

4. Part A: $8 \times \frac{1}{2} = \frac{8}{2}$ = 4 pounds of fruit salad
   Part B: The family drank a total of 4,100 milliliters of iced tea.
   3 liters = 3,000 milliliters: 3,000 + 1,600 = 4,600
   0.5 liters = 500 milliliters: 4,600 – 500 = 4,100
   Part C: The area of the tablecloth is 40 square feet.
   Length of table = 6 ft + 2 ft = 8 ft
   Width of table = 3 ft + 2 ft = 5 ft
   $8 \times 5 = 40$ square feet
   Part D: They arrived at the park at 9:40 A.M.
   Part E: The total angle measurement of the reclining chair is 133°.
   The upright angle is a right angle, so it is 90 degrees.
   The reclining position is another 43 degrees. 90 + 43 = 133.

5. A. T
   B. F (should not include 1 minute, so the total is 3 minutes)
   C. F

Chapter Checklist for Standards

If you answered the questions correctly, you are on your way toward mastering the concepts and skills for this standard.

Standard	Concept/Skill	Questions
4.MD.1	I know relative sizes of units within one system of units.	1, 2, 3
4.MD.2	I use the four operations to solve word problems using measurements.	2, 4
4.MD.3	I apply area and perimeter formulas for rectangles.	4
4.MD.4	I make a line plot/solve problems involving a line plot.	5
4.MD.5-6-7	I measure and recognize angle measures as additive.	4

CHAPTER 7

Geometry (G)

When we study geometry we study shapes and figures. In grade 4, these shapes and figures are located on a plane, or flat surface.

VOCABULARY

Point: a single location that can be identified with a letter

- *A* is known as point *A*

Line: when points are combined and extended in a straight direction, they form a line; a line is straight and extends in both directions

Line segment: part of a line with points that show where it starts and stops

Ray: part of a line that stops in one direction and extends in the other direction

We identify points, line segments, and rays on lines

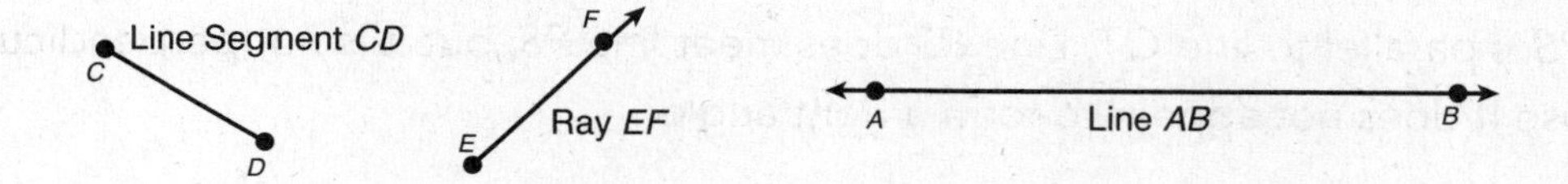

Angles: formed by 2 rays with a common endpoint known as the vertex; angles are identified with 3 letters for the 3 points or by 1 letter using the vertex; types of angles are identified by the angle measures

Acute angles: have a measure of less than 90 degrees

Right angles: have a measure of exactly 90 degrees

Obtuse angles: have a measure larger than 90 and less than 180 degrees

Straight angles: have a measure of exactly 180 degrees

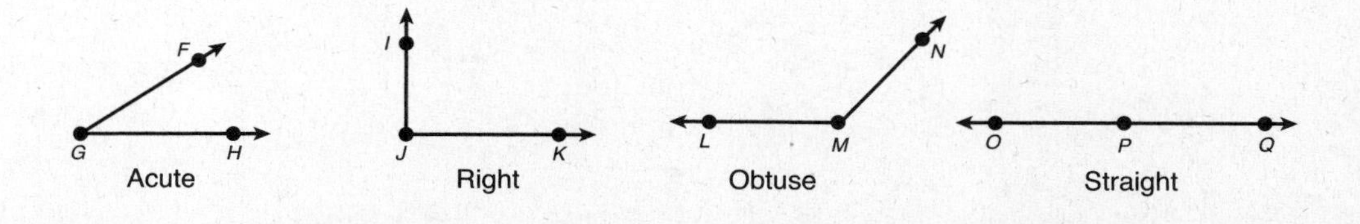

Parallel and Perpendicular Lines

We often see lines, line segments, and rays in pairs or groups. We describe how they relate to each other's location by classifying them as parallel or perpendicular.

Parallel lines, line segments, and rays are the same distance apart and extend in the same direction. Parallel lines will never meet, or cross.

Perpendicular lines, line segments, and rays meet and form right angles measuring exactly 90 degrees.

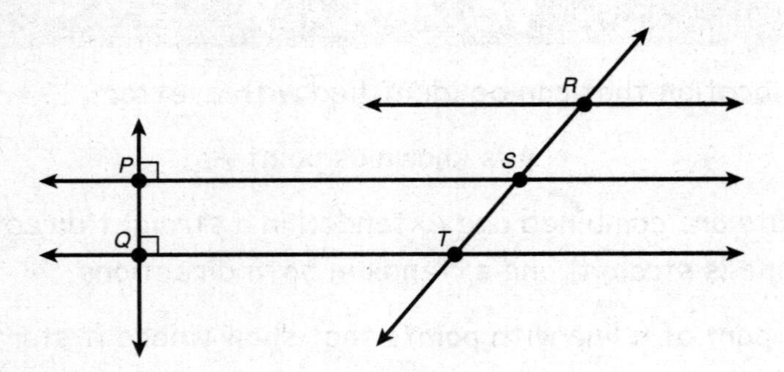

In the example shown, line *PS* is perpendicular to line *PQ*. You can tell by the right-angle symbol located on point *P*. Line *QT* is also perpendicular to line *PQ*. Line *PS* is parallel to line *QT*. Line *RS* does meet line *PS*, but it is not perpendicular because it does not appear to form a right angle.

Polygon Review

Line segments can form the sides of shapes known as polygons.

Definition

Flat, closed shapes with three or more straight sides that meet and form angles

We can classify polygons in different ways.

Number of Sides

Triangles: three sides
Quadrilaterals: four sides
Pentagons: five sides
Hexagons: six sides
Octagons: eight sides

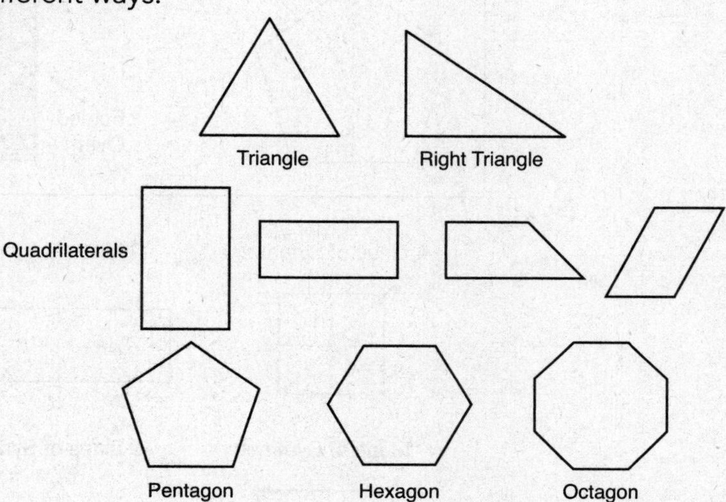

Parallel Sides and Perpendicular Sides

Parallelograms: quadrilateral with two pairs of parallel sides
Rectangles: parallelograms with perpendicular sides

Parallelograms

Rectangles

Equal Sides

Rhombus: parallelogram with four equal sides
Square: rhombus with perpendicular sides

Rhombus

Square

Lines of Symmetry

Some flat shapes have lines of symmetry. Shapes with a line of symmetry, when folded along that line, have matching parts. Lines of symmetry can appear in different directions:

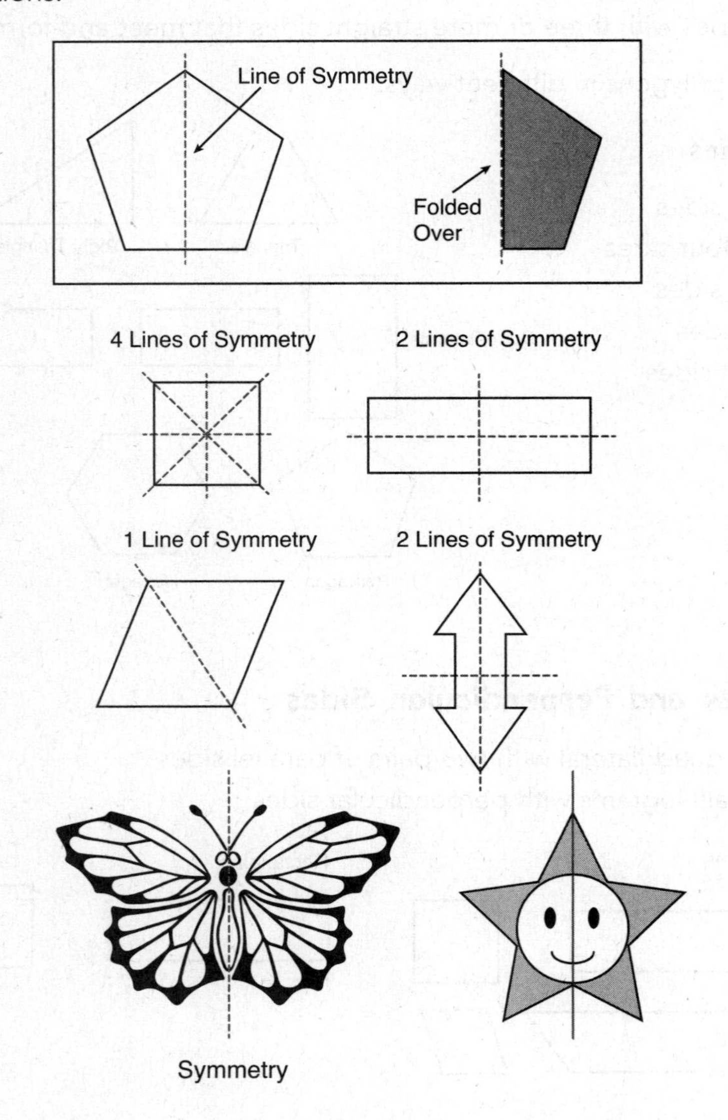

Letters and pictures can also have lines of symmetry:

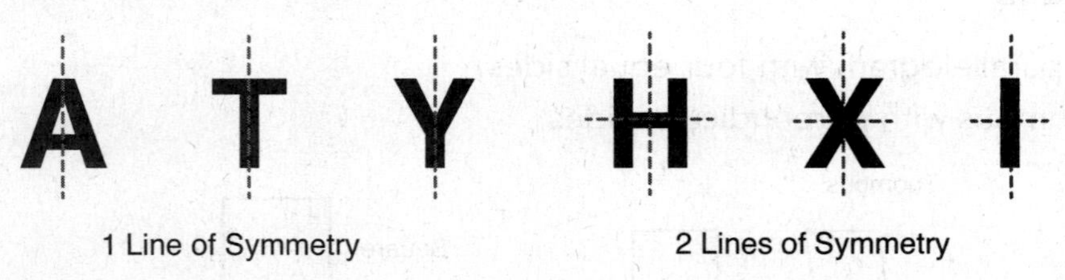

Geometry Practice Problems

1. Use the diagram to complete the table:

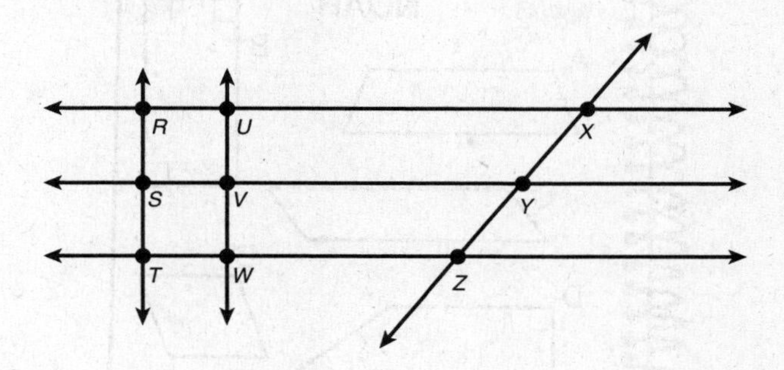

Place a check to correctly complete the table to describe the pair of lines.

Pair	Perpendicular	Parallel	Neither
A. Line *RT* and line *UW*			
B. Line *RX* and line *SY*			
C. Line *SV* and line *ST*			
D. Line *TZ* and line *XZ*			
E. Line *ST* and line *TW*			

2. Noah placed some stickers on his notebook. The stickers are all polygons.

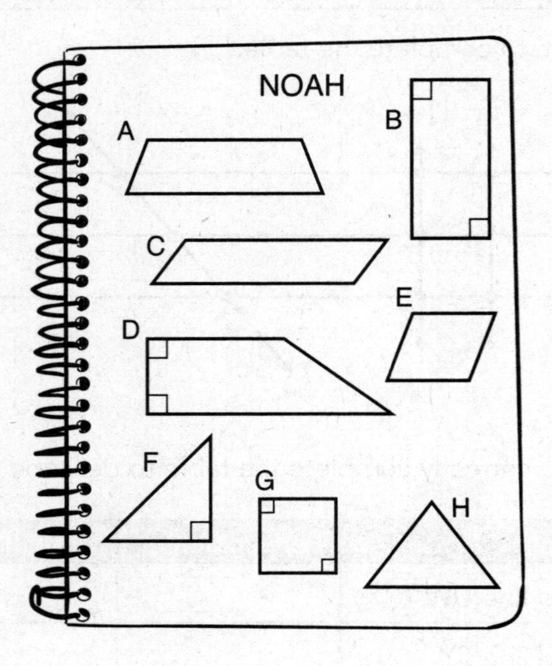

PART A

Which FOUR shapes on Noah's notebook have two pairs of parallel sides?

☐ A. Shape A
☐ B. Shape B
☐ C. Shape C
☐ D. Shape D
☐ E. Shape E
☐ F. Shape F
☐ G. Shape G
☐ H. Shape H

PART B

Noah said that shapes F and H are right triangles. Is Noah correct?
Why or why not?

PART C

Place the letter of the shape in the box with the title that correctly describes the shape. Some letters will be used more than once or not at all.

Shapes with Only Right Angles	Shapes with Only Acute and Obtuse Angles	Shapes with Right, Acute, and Obtuse Angles

3. Shapes A–F are shown below. Check the boxes to correctly complete the statements in the table to describe the number of lines of symmetry for each of the shapes. Check only one box in each row.

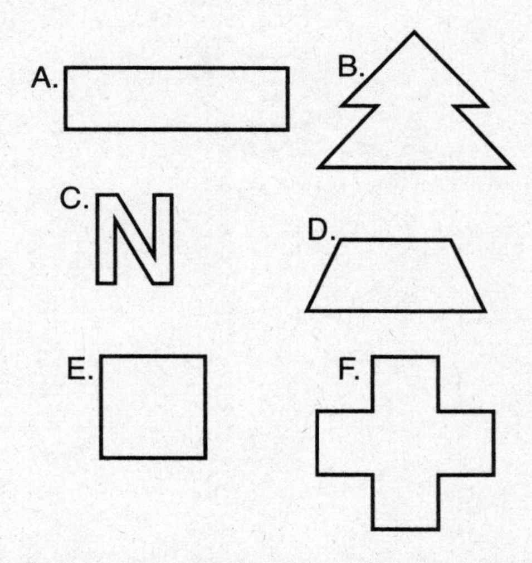

Shape	No Lines of Symmetry	Exactly One Line of Symmetry	Exactly Two Lines of Symmetry	More Than Two Lines of Symmetry
A.				
B.				
C.				
D.				
E.				
F.				

4. PART A

Construct a figure using the following descriptions:

- The figure *ABCD* is not a parallelogram.
- Side *BC* is parallel to side *AD*.
- Angles *B* and *C* are acute angles and angles *A* and *D* are obtuse angles.

PART B

Is figure *ABDC* a rectangle? Why or why not?

Answers to Geometry Practice Problems

1. A. Parallel
   B. Parallel
   C. Perpendicular
   D. Neither
   E. Perpendicular

2. Part A: **B, C, E, G**
   Part B: No. F is a right triangle but H is not because it does not have any right angles.
   Part C: Only right angles: B and G
   Only acute and obtuse: A, C, and E
   Right, acute, and obtuse: D

3. A. Two lines of symmetry
   B. Exactly one line of symmetry
   C. No lines of symmetry
   D. Exactly one line of symmetry
   E. More than two lines of symmetry
   F. More than two lines of symmetry

4. Part A: Correct figure:

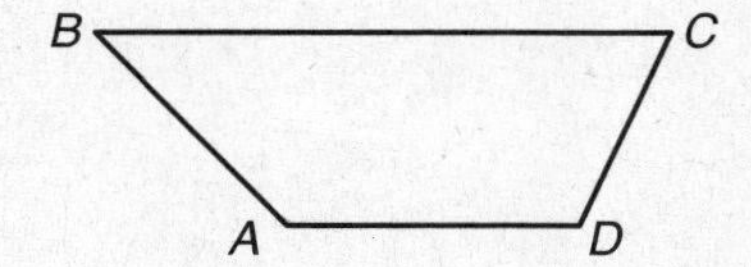

   Part B: The figure is not a rectangle because it does not have any right angles. *ABCD* is a trapezoid.

Chapter Checklist for Standards

If you answered the questions correctly, you are on your way toward mastering the concepts and skills for this standard.

Standard	Concept/Skill	Questions
4.G.1	I draw points, lines, rays, angles, and pairs of lines. I identify these in two-dimensional figures.	1, 4
4.G.2	I classify two-dimensional figures based on the presence or absence of pairs of lines, or angles of a specified size.	1, 2
4.G.3	I recognize a line of symmetry and identify line-symmetric figures.	3

CHAPTER 8

Common Core Review

Common Core Review Practice Problems

1. The chart below shows the points scored from the last game of the junior league football teams.

Team	Points	Team	Points
Wildcats	21	Tigers	7
Bears	9	Cardinals	28
Bulldogs	14	Eagles	3

Which FOUR Statements are true?

☐ A. The Bulldogs scored 5 times as many points as the Bears.

☐ B. The Eagles scored 3 times as many points as the Bears.

☐ C. The Wildcats scored 3 times as many points as the Tigers.

☐ D. The Cardinals scored 4 times as many points as the Tigers.

☐ E. The Wildcats scored 7 times as many points as the Eagles.

☐ F. The Bulldogs scored 2 times as many points as the Tigers.

2. Frank made a chart to find some of the prime and composite factors of 36. Place the numbers in the boxes to correctly complete Frank's chart. Use each number once or not at all.

1	2	3	4	5	6	8	9	12	16	18	36

Factors of 36

Prime Factors	Composite Factors

3. Jaden wants to compare the values of the three coins shown:

Place the responses below on the lines to make each statement true. You may use the responses more than once.

5 more	10 more	5 times more	10 times more

A. The value of 1 dime is __________ than the value of 1 nickel.

B. The value of 1 quarter is __________ than the value of 1 nickel.

C. The value of 1 quarter is __________ than the value of 2 dimes.

D. The value of 2 quarters is __________ than the value of 2 nickels.

E. The value of 2 quarters is __________ than the value of 4 dimes.

F. The value of 2 quarters is __________ than the value of 1 nickel.

4. Mackenzie made a number pattern using the rule "subtract 7." Which pattern is Mackenzie's number pattern?

O A. 1, 7, 13, 19, 25, 31
O B. 77, 70, 64, 59, 55, 52
O C. 7, 14, 21, 28, 35, 42
O D. 60, 53, 46, 39, 32

5. Keilani wrote a number in word form:

Twenty-eight thousand, seven hundred fifteen

Keilani wants to write a new number in standard form using the following rules:

- The value of the digits 1, 5, 7, and 8 have a value that is 10 times more than the same digits in Keilani's first number written in word form.
- The number is an even number.

Use the digits shown to write the standard form of Keilani's new number. Use each digit once.

Digits to use: 1 2 3 5 7 8

Keilani's new number: ___ ___ ___ , ___ ___ ___

6. Lily solved the equation 4,198 + 2,736 = ?

First she found the actual sum. Then she rounded the numbers to the nearest hundred and estimated the sum to make sure that her first answer was reasonable.

Which TWO values are Lily's actual and estimated sums?

☐ A. 6,700
☐ B. 6,900
☐ C. 6,824
☐ D. 6,834
☐ E. 6,934
☐ F. 7,000

7. Steffany solved some multiplication problems by rounding the factors to find the estimated product:

$$? \times ? = 2{,}400$$

Which of the following could be the problems that Steffany solved?
Circle Y (yes) or N (no) for each.

A. $31 \times 79 = ?$ Y N
B. $4 \times 620 = ?$ Y N
C. $60 \times 399 = ?$ Y N
D. $21 \times 1{,}205 = ?$ Y N
E. $82 \times 27 = ?$ Y N
F. $11 \times 22 = ?$ Y N

8. Solve the following division problems by placing the correct quotient on the matching division problem. Use each quotient once or not at all.

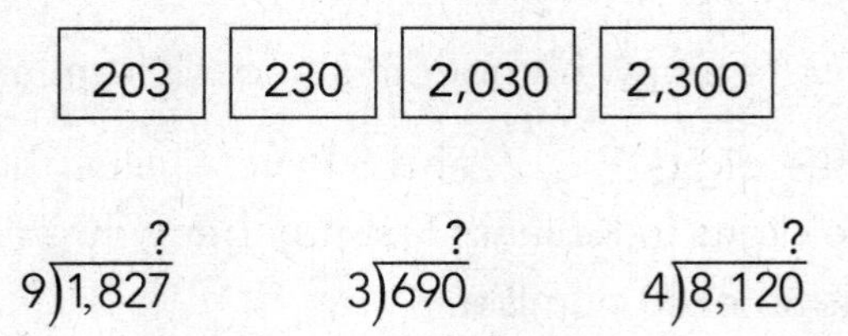

9. Compare the fractions by using the fraction chart shown:

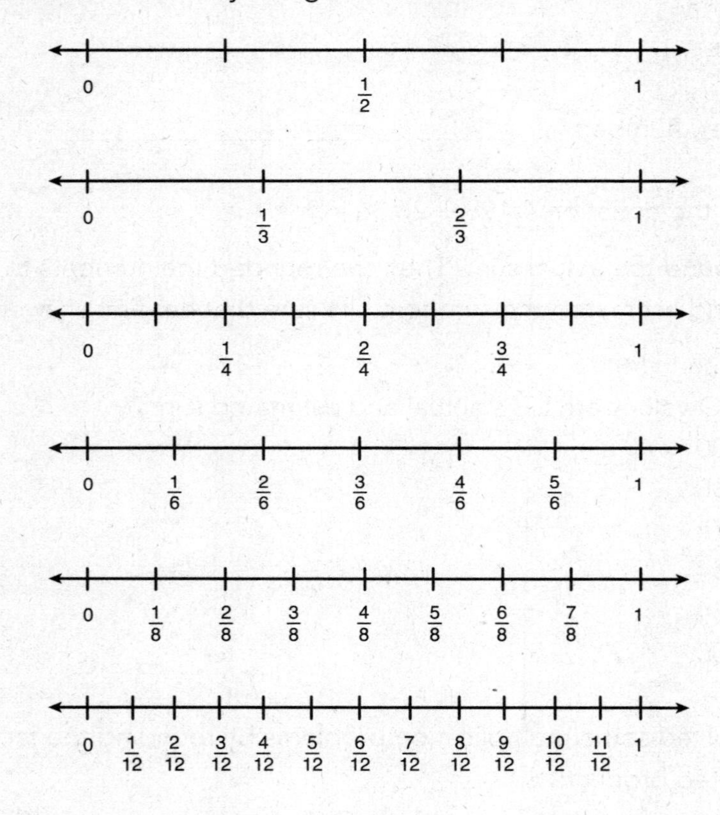

Place the correct sign in the box to correctly complete each number sentence. Each sign can be used more than once.

< = >

$\frac{1}{2}$ ☐ $\frac{5}{8}$ $\frac{2}{3}$ ☐ $\frac{3}{4}$ $\frac{1}{4}$ ☐ $\frac{3}{12}$

$\frac{2}{6}$ ☐ $\frac{2}{8}$ $\frac{6}{6}$ ☐ $\frac{1}{1}$ $\frac{5}{12}$ ☐ $\frac{7}{12}$

10. There are two quart containers below. Each has an amount of water, in quarts, shown.

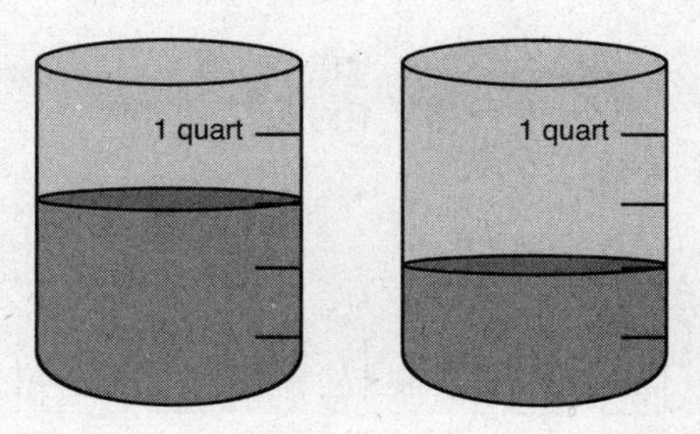

Which of the following shows the total amount of water, in quarts, of both containers?

O A. $\frac{5}{8}$

O B. $\frac{5}{5}$

O C. $1\frac{1}{5}$

O D. $1\frac{1}{4}$

11. Point *X* is shown on the number line.

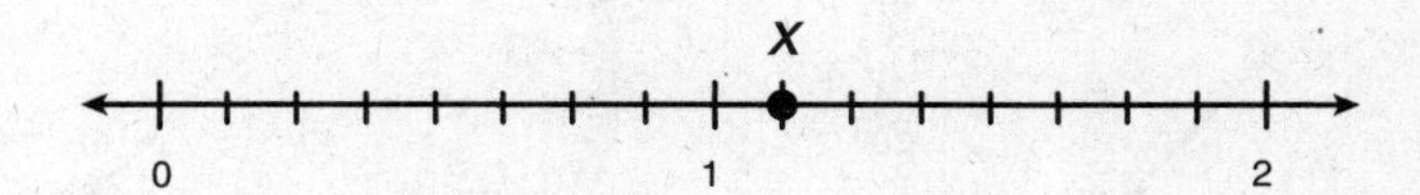

Which THREE equations can be used to show how to get from 0 to point *X* on the number line?

☐ A. $\frac{5}{10}+\frac{5}{10}=?$

☐ B. $3\times\frac{3}{8}=\frac{9}{8}=1\frac{1}{8}$

☐ C. $10\times\frac{1}{10}=?$

☐ D. $\frac{5}{8}+\frac{4}{8}=\frac{9}{8}=1\frac{1}{8}$

☐ E. $3\times\frac{3}{9}=?$

☐ F. $9\times\frac{1}{8}=\frac{9}{8}=1\frac{1}{8}$

☐ G. $\frac{5}{9}+\frac{4}{9}=?$

12. Solve the equation by renaming the fractions as decimals. Place the digits in the boxes shown to correctly solve. Use the digits more than once or not at all.

$$\frac{3}{10}+\frac{48}{100}=?$$

Digits to be used:

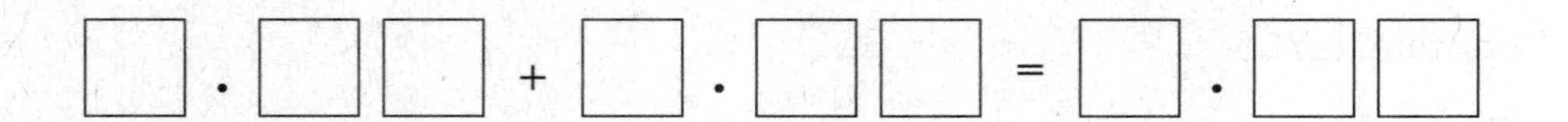

13. Shown below are two scales that show two different mass measurements in kilograms.

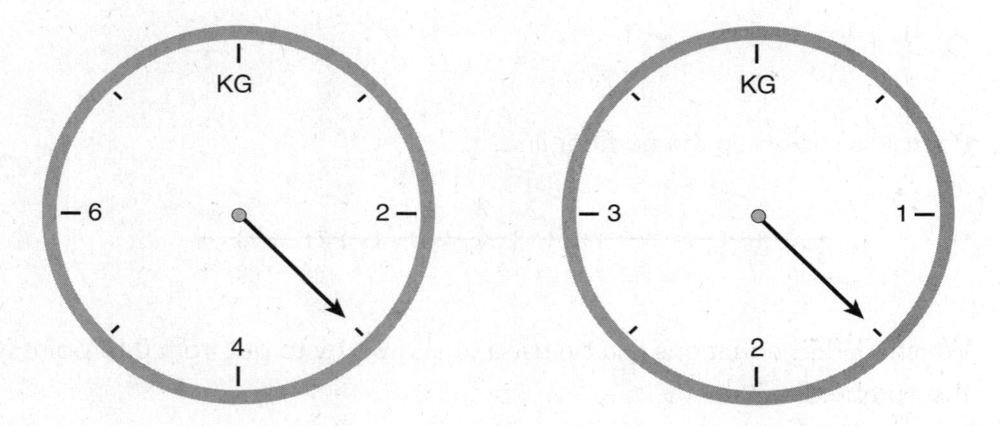

Which TWO values show the correct combined mass of the two scales shown?

☐ A. 3.5 kilograms
☐ B. 4.5 kilograms
☐ C. 6 kilograms
☐ D. 3,500 grams
☐ E. 4,500 grams
☐ F. 6,000 grams

14. The square shown below has a side length of 6 inches.

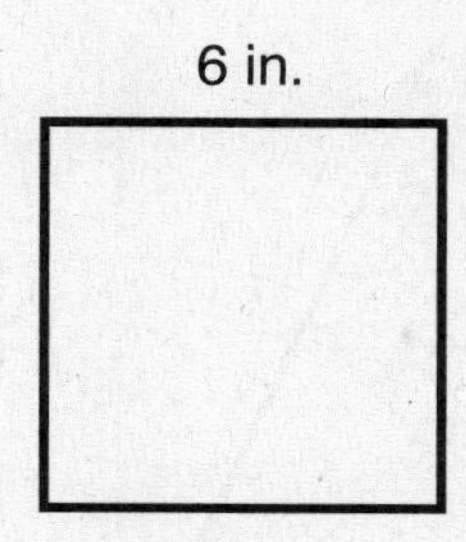

Place the digits and words on the lines to correctly complete each statement.

6	12	24	36	inches	sq. inches

The perimeter of the square is ____ ____ .

The area of the square is ____ ____.

15. The line plot below shows the weights, in pounds, of 7 bags of rocks.

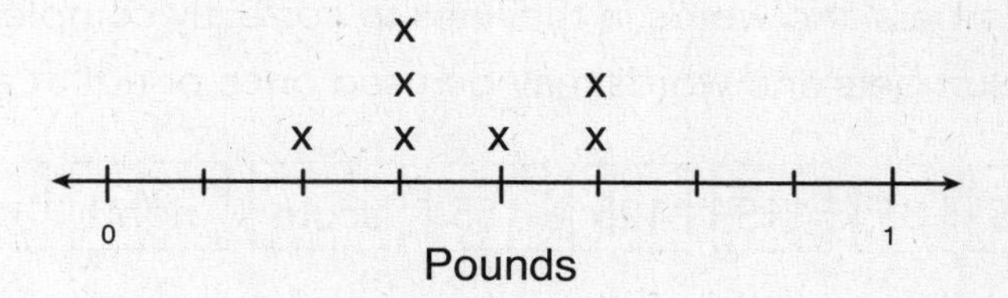

After the weight of the 8th bag is placed on the line plot, the difference between the greatest and least weights is $\frac{5}{8}$ pound. Which of the following shows the weight, in pounds, of the 8th bag of rocks?

O A. $\frac{2}{8}$

O B. $\frac{3}{8}$

O C. $\frac{5}{8}$

O D. $\frac{7}{8}$

16. Ben drew some angles as shown below:

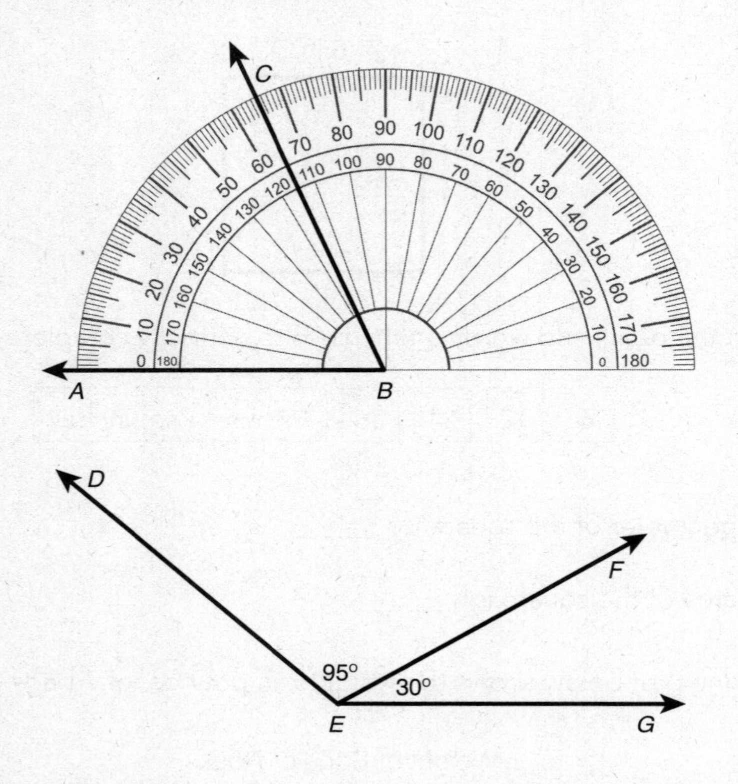

Place the numbers and words in the lines to correctly complete each statement. Numbers and words may be used once or not at all.

65	75	85	115	125	155	acute	right	obtuse

Angle *ABC* measures _______ degrees and is a/an _______ angle.

Angle *DEF* measures 95 degrees and is a/an _______ angle.

Angle *DEG* measures _______ degrees and is a/an _______ angle.

17. Stephanie drew some shapes as shown below:

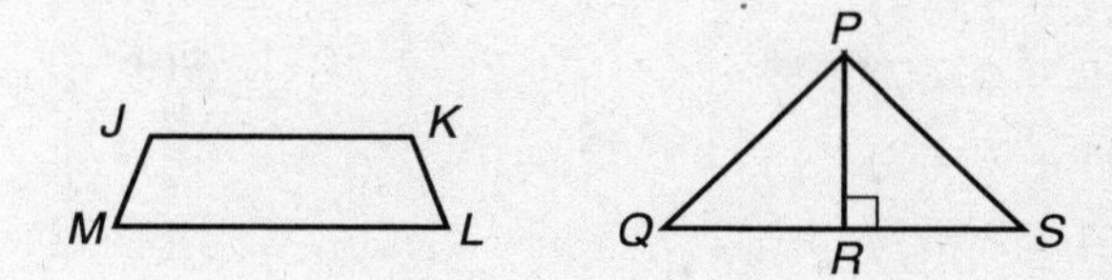

Which of the following statements are true? Circle T (true) or F (false) to correctly complete each statement.

A. Line segment *JK* is parallel to line segment *LM*. T F

B. Line segment *JK* is perpendicular to line segment *LM*. T F

C. Shape *JKLM* is a parallelogram. T F

D. Shape *JKLM* is a trapezoid. T F

E. Shape *PRS* is a right triangle. T F

F. Shape *PSQ* is a right triangle. T F

18. Place the letters of the shapes in the boxes to correctly complete the chart.

A. B. C. D.

Shapes with No Lines of Symmetry	Shapes with Exactly One Line of Symmetry	Shapes with More Than One Line of Symmetry

Answers to Common Core Review Practice Problems

1. ***C*** Correct: $(21 \div 3 = 7)$
   D Correct: $(28 \div 4 = 7)$
   E Correct: $(21 \div 7 = 3)$
   F Correct: $(14 \div 2 = 7)$

2. Prime: 2, 3
   Composite: 4, 6, 9, 12, 18, 36

3. A. The value of 1 dime is 5 more than the value of 1 nickel.
   $(10 - 5 = 5)$
   B. The value of 1 quarter is 5 times more than the value of 1 nickel.
   $(25 \div 5 = 5)$
   C. The value of 1 quarter is 5 more than the value of 2 dimes.
   $(25 - 5 = 20)$
   D. The value of 2 quarters is 5 times more than the value of 2 nickels.
   $(50 \div 5 = 10)$
   E. The value of 2 quarters is 10 more than the value of 4 dimes.
   $(50 - 10 = 40)$
   F. The value of 2 quarters is 10 times more than the value of 1 nickel.
   $(50 \div 10 = 5)$

4. **D** $60 - 7 = 53$; $53 - 7 = 46$; $46 - 7 = 39$; $39 - 7 = 32$

5. 387,152 (The digits 1, 5, 7, and 8 are moved one place to the left, and the 2 is in the ones place because the number is even, leaving the 3 for the hundred thousands place.)

6. **B** Correct: 6,900 (estimated answer: 4,200 + 2,700)
   E Correct: 6,934 (actual answer)

7. A. Y (30×80)
   B. Y (4×600)
   C. N
   D. N
   E. Y (80×30)
   F. N

8. $\begin{array}{r}203\\9\overline{)1,827}\end{array}$ $\begin{array}{r}230\\3\overline{)690}\end{array}$ $\begin{array}{r}2,030\\4\overline{)8,120}\end{array}$

9. $\frac{1}{2} < \frac{5}{8}$; $\frac{2}{3} < \frac{3}{4}$; $\frac{1}{4} = \frac{3}{12}$; $\frac{2}{6} > \frac{2}{8}$; $\frac{6}{6} = \frac{1}{1}$; $\frac{5}{12} < \frac{7}{12}$

10. **D** $\left(\frac{3}{4} + \frac{1}{2} = \frac{3}{4} + \frac{2}{4} = \frac{5}{4} = 1\frac{1}{4}\right)$

11. **B** Correct: $3 \times \frac{3}{8} = \frac{9}{8} = 1\frac{1}{8}$

    D Correct: $\frac{5}{8} + \frac{4}{8} = \frac{9}{8} = 1\frac{1}{8}$

    F Correct: $9 \times \frac{1}{8} = \frac{9}{8} = 1\frac{1}{8}$

12. 0.30 + 0.48 = 0.78

13. **B** Correct: 3 kg + 1.5 kg = 4.5 kg
    E Correct: 3,000 g + 1,500 g = 4,500 g

14. Perimeter = 24 inches (6 + 6 + 6 + 6)
    Area = 36 square inches (6 × 6)

15 **D** $\frac{2}{8} + \frac{5}{8} = \frac{7}{8}$

16. Angle *ABC* measures 65 degrees and is an acute angle.
    Angle *DEF* measures 95 degrees and is an obtuse angle.
    Angle *DEG* measures 125 degrees and is an obtuse angle.

17. A. T
    B. F
    C. F
    D. T
    E. T
    F. F

18. No lines of symmetry: B
    Exactly one line of symmetry: D
    More than one line of symmetry: A and C

Checklist for Standards

If you answered the questions correctly, you are on your way toward mastering the concepts and skills for the Grade 4 Standards.

Standard	Question
4.OA.1	1
4.OA.2	2
4.OA.4	3
4.OA.5	4
4.NBT.1	5
4.NBT.3	6,7
4.NBT.6	8
4.NF.2	9
4.NF.3	10, 11
4.NF.5	12
4.MD.1	13
4.MD.2	10
4.MD.3	14
4.MD.4	15
4.MD.5	16
4.G.1-2	17
4.G.3	18

Practice Test 1

IMPORTANT NOTE

Barron's has made every effort to create sample tests that accurately reflect the PARCC Assessment. However, the tests are constantly changing. The following two tests differ in length and content, but each will still provide a strong framework for fourth-grade students preparing for the assessment. Be sure to consult *http://www.parcconline.org* for all the latest testing information.

Session 1 (50 minutes)

You may NOT use a calculator for these questions. The estimated time for completing this session is 50 minutes; however, you may be permitted more time if you need it. Follow all directions to complete each question.

1. Which number is 7 times as many as the number 9?

- O A. 16
- O B. 49
- O C. 56
- O D. 63

2. In the number 57,680, the value of the digit 7 is 10 times more than the value of the digit 7 in which THREE numbers?

- ☐ A. 65,780
- ☐ B. 65,870
- ☐ C. 75,850
- ☐ D. 75,760
- ☐ E. 86,750
- ☐ F. 87,650

3. Place the correct symbol in the box to make the number sentences true.

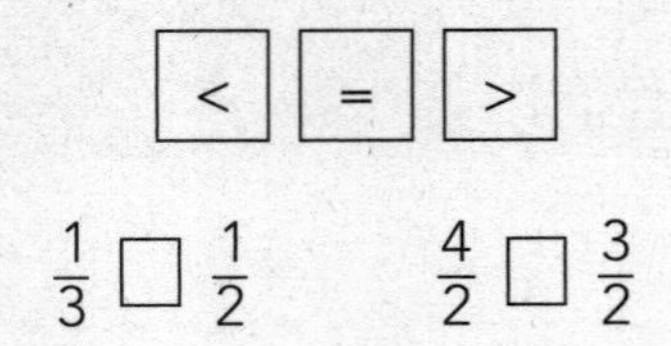

$\frac{1}{3} \square \frac{1}{2}$ $\frac{4}{2} \square \frac{3}{2}$

4. The table below shows the colors and number of sweatshirts ordered so far by the principal of the Center St. School.

Color	Number of Sweatshirts
Red	187
Gray	215
Black	246

PART A

Find the total number of sweatshirts ordered so far. ___________

PART B

The principal ordered some white sweatshirts. The new total of sweatshirts is 875. Write an equation to find the number of white sweatshirts ordered. Solve your equation.

PART C

The total number of sweatshirts in the 4 colors are being placed in boxes of 8. How many boxes will be needed for the 875 sweatshirts?

5. Point Q is shown on the number line below:

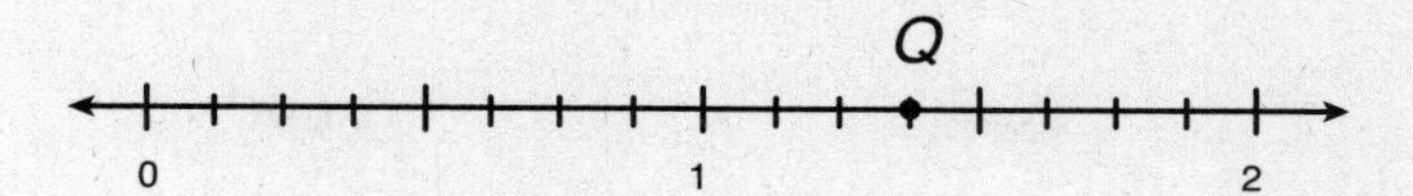

Place the numbers in the boxes to correctly complete an equation that can be used to find the location of point Q as an improper number and as a mixed number. Numbers can be used once or not at all.

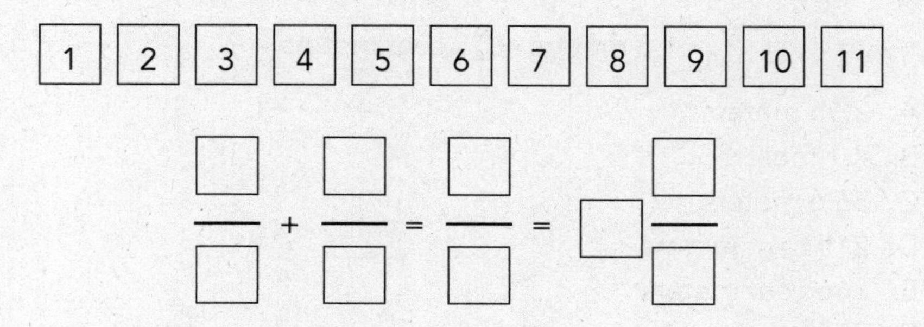

6. MaryJane solved the multiplication problem below incorrectly.

$$\begin{array}{r} 1{,}638 \\ \times \quad 4 \\ \hline 4{,}422 \end{array}$$

PART A

Explain why MaryJane's multiplication problem is incorrect.

PART B

Solve MaryJane's multiplication problem correctly.

PART C

Write an equation using a different operation that could be used to check your multiplication problem.

7. Solve: $0.04 = \frac{?}{?}$

8. Which TWO values correctly solve the equation

$$1.6 \text{ meters} + 55 \text{ centimeters} = ?$$

☐ A. 2.15 meters
☐ B. 7.1 meters
☐ C. 56.6 meters
☐ D. 215 centimeters
☐ E. 566 centimeters
☐ F. 710 centimeters

Session 2 (50 minutes)

You may NOT use a calculator for these questions. The estimated time for completing this session is 50 minutes; however, you may be permitted more time if you need it.

1. Ali wants to use the array to model the equation $5 \times 5 = ?$ Type 2: 3 points

PART A

Does Ali's array correctly model the equation shown? Why or why not?

PART B

Create an array model for the equation $3 \times 8 = 24$.

2. Which THREE number pairs are factor pairs of 32?

☐ A. 1×32
☐ B. 2×16
☐ C. 3×12
☐ D. 4×8
☐ E. 5×6

3. Carley started to write a number pattern, using one operation, as shown below:

3, ____, 11, ____, ____, 23

The rule of Carley's number pattern is ________ ________ .

Use the numbers shown below to correctly complete the pattern and the statement above:

3	4	7	8	12	15	16	19	20	+	–	×	÷

4. The population of the city where Mr. Allan lives is shown below:

One hundred forty thousand, three hundred five

Which of the following is the standard form?

O A. 104,305
O B. 104,350
O C. 140,305
O D. 140,350

5. Aidan wrote some numbers that he wants to compare and round.

PART A

Circle the symbol that correctly compares the number pairs shown:

155,701 < = > 154,810

174,618 < = > 173,627

PART B

Place the numbers in the chart to correctly round them to the nearest ten thousand.

Rounds to 150,000	Rounds to 160,000	Rounds to 170,000	Rounds to 180,000

6. Solve: $36 \times 85 =$ _______

7. Rose measured the angle formed by her scissors. Rose continued to open the scissors, until the angle measurement was 135 degrees. How many more degrees did Rose open the scissors?

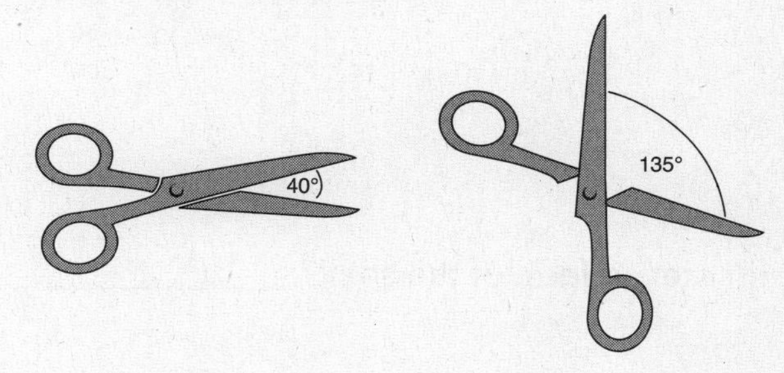

O A. 5°
O B. 85°
O C. 95°
O D. 175°

8. Ana has a shelf in her room. The top view of the shelf is shown below:

4 feet

$\frac{3}{4}$ foot

PART A

Which TWO equations could be used to find the area, in square feet, of the top view of Ana's shelf?

☐ A. $3 \times \frac{1}{4} = ?$

☐ B. $4 \times \frac{1}{4} = ?$

☐ C. $3 \times \frac{3}{4} = ?$

☐ D. $4 \times \frac{3}{4} = ?$

☐ E. $12 \times \frac{1}{4} = ?$

☐ F. $12 \times \frac{3}{4} = ?$

What is the area, in square feet, of the shelf? ______________________

PART B

Write an equation that could be used to find the perimeter of Ana's shelf.

What is the perimeter, in feet, of the shelf? ______________________

9. Alex made a shape using 6 equal-sized triangles:

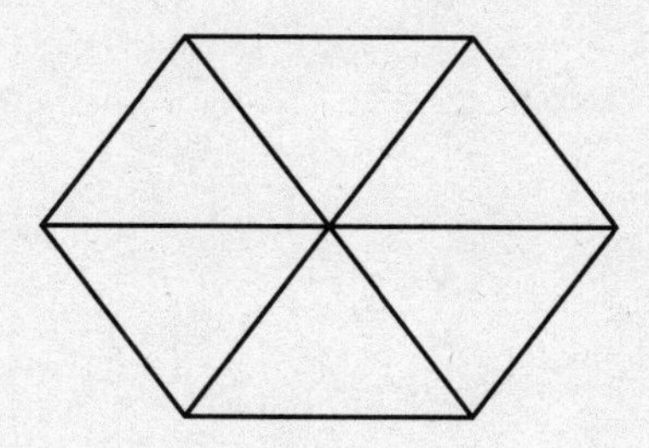

PART A

Alex's shape is 1 whole. Fill in the blanks to correctly complete the statements.

A. The fraction $\frac{1}{6}$ could be shown using ______ triangle(s).

B. The fraction $\frac{1}{3}$ could be shown using ______ triangle(s).

C. The fraction $\frac{1}{2}$ could be shown using ______ triangle(s).

D. The fraction $1\frac{1}{2}$ could be shown using ______ triangle(s).

PART B

Place the fractions from Part 1 on the number line to show their location. Label the fractions using the points as shown:

Point A: $\frac{1}{6}$ Point B: $\frac{1}{3}$ Point C: $\frac{1}{2}$ Point D: $1\frac{1}{2}$

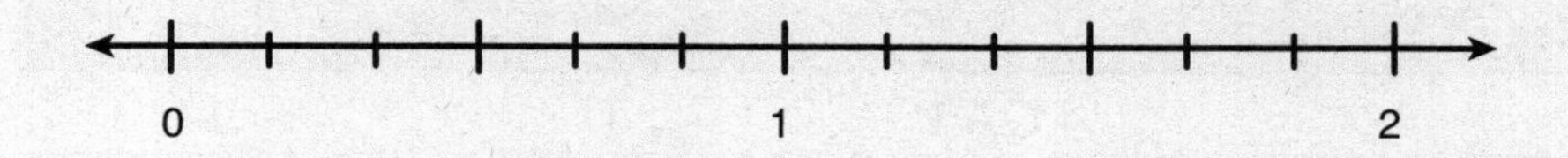

Answers to Practice Test 1

Session 1

1. **D** 63

2. **A** Correct: 65,780
   D Correct: 75,760
   E Correct: 86,750

3. $\frac{1}{3} < \frac{1}{2}$; $\frac{4}{2} > \frac{3}{2}$

4. Part A: 648 sweatshirts
   Part B: $875 - 648 = 227$ 227 white sweatshirts were ordered.
   Part C: $875 \div 8 = 109$ R3 110 boxes are needed.

5. $\frac{6}{8} + \frac{5}{8} = \frac{11}{8} = 1\frac{3}{8}$ or any combination that equals $\frac{11}{8}$.

6. Part A: MaryJane did not carry to the next place.
   Part B: $4 \times 1{,}638 = 6{,}552$
   Part C: $6{,}552 \div 4 = 1{,}638$

7. $0.04 = \frac{4}{100}$

8. **A** Correct: 2.15 meters
   D Correct: 215 centimeters

Checklist for Standards

Type and point value for each question.

Question	Standard	Type	Point Value
1	4.OA.1	I	1
2	4.NBT.1	I	1
3	4.NF.2	I	1
4	4.OA.3	II	4
5	4.NF.3	II	1
6	4.NBT.5	II	3
7	4.NF.6	I	1
8	4.MD.1-2	I	1

Session 2

1. Part A: Ali's array is not correct. The array shows 2 rows of 5, or 5 + 5, or 10. Ali would need 5 rows of 5.
   Part B: An array similar to 3 rows of 8, or 8 rows of 3.
2. **A** Correct: 1×32
   B Correct: 2×16
   D Correct: 4×8
3. 7, 15, 19
   Rule: + 4
4. ***C*** 140,305
5. Part A: 155,701 > 154,810
   174,618 > 173,627
   Part B: Rounds to 150,000: 154,810
   Rounds to 160,000: 155,701
   Rounds to 170,000: 173,627 and 174,618
   Rounds to 180,000: None
6. $36 \times 85 = 3{,}060$
7. ***C*** 95°
8. Part A: **D** Correct: $4 \times \frac{3}{4} = ?$

   E Correct: $12 \times \frac{1}{4} = ?$

   Area = 3 square feet or $\frac{12}{4}$ square feet

   Part B: Equation: $4 + \frac{3}{4} + 4 + \frac{3}{4} = ?$

   Perimeter = either $8\frac{6}{4}$, $9\frac{2}{4}$, $9\frac{1}{2}$ feet.

9. Part A: A. The fraction $\frac{1}{6}$ could be shown using 1 triangle.

B. The fraction $\frac{1}{3}$ could be shown using 2 triangles.

C. The fraction $\frac{1}{2}$ could be shown using 3 triangles.

D. The fraction $1\frac{1}{2}$ could be shown using 9 triangles.

Part B:

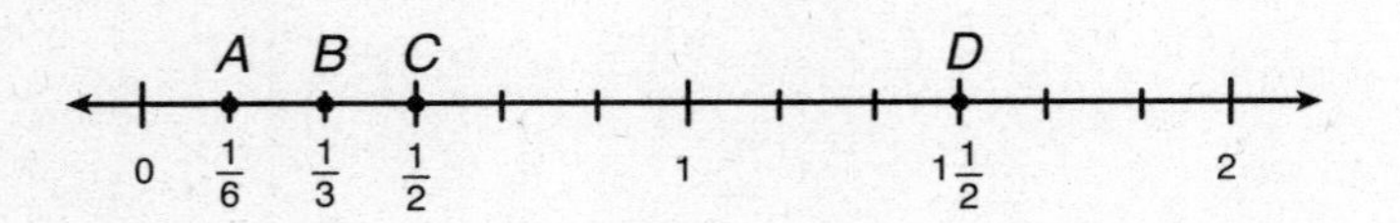

Checklist for Standards

Type and point value for each question.

Question	Standard	Type	Point Value
1	4.OA.2	II	3
2	4.OA.4	I	1
3	4.OA.5	I	2
4	4.NBT.2	I	1
5	4.NBT.2	I	2
6	4.NBT.5	I	1
7	4.MD.3 and 4.NF.3	III	4
8	4.MD.7	I	1
9	4.NF.1	III	6

Practice Test 2

Session 1 (55 minutes)

You may NOT use a calculator for these questions. The estimated time for completing this session is 55 minutes; however, you may be permitted more time if you need it. Follow all directions to complete each question.

1. Which number is 120 less than the product of 215 and 38?

O A. 3,610
O B. 4,560
O C. 8,050
O D. 8,170

2. Place the numbers in the box to correctly complete the statement. Some numbers will not be used.

48	6	40	7	8	54

The number ______ is 6 times more than the number ______.

3. Which is true about the number 9? Choose THREE statements that are correct.

☐ A. The number is composite.
☐ B. The number 3 is a factor.
☐ C. The number is prime.
☐ D. The number 36 is a multiple.
☐ E. The number 4 is a factor.
☐ F. The number 64 is a multiple.

4. Which number in written form correctly completes the number sentence shown?

$$100{,}000 + 6{,}000 + 400 + 90 + 6 > ?$$

- O A. One hundred fifty thousand, four hundred ninety-three
- O B. One hundred sixty thousand, nine hundred five
- O C. One hundred five thousand, four hundred ninety-seven
- O D. One hundred six thousand, five hundred ninety-four

5. A pattern has been started as shown:

_____ , 77, 66, _____ , 44, _____

PART A

Circle the symbol and number to correctly complete the statement:

The rule of the pattern is _______ _______ .

+	6
−	7
÷	11

PART B

What is the first number that starts the pattern? ________

6. Place the numbers in the box to correctly complete the addition problem shown below. Some numbers may be used more than once or not at all.

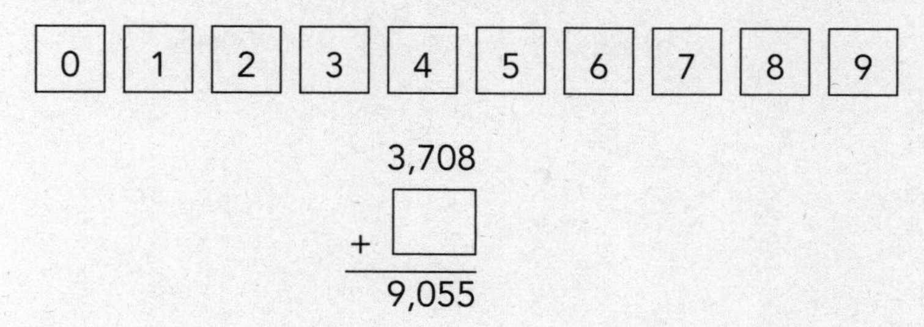

7. Circle the numbers to correctly complete the equation shown:

5,618 ÷ 7 = ______ R ______

8,200	0
820	1
802	4

8. Use the fractions shown to complete the chart. Some fractions will not be used.

$\frac{3}{6}$ $\frac{4}{6}$ $\frac{5}{12}$ $\frac{6}{12}$

Fraction > $\frac{1}{2}$	Fraction < $\frac{1}{2}$

9. Which value correctly completes the equation shown below?

$$3\frac{3}{4} + ? = 5\frac{1}{4}$$

- O A. $\frac{2}{4}$
- O B. $\frac{3}{4}$
- O C. $1\frac{1}{4}$
- O D. $1\frac{2}{4}$

10. Angle *GHI* is shown below.

PART A

Which shows the measure of angle *GHI*?

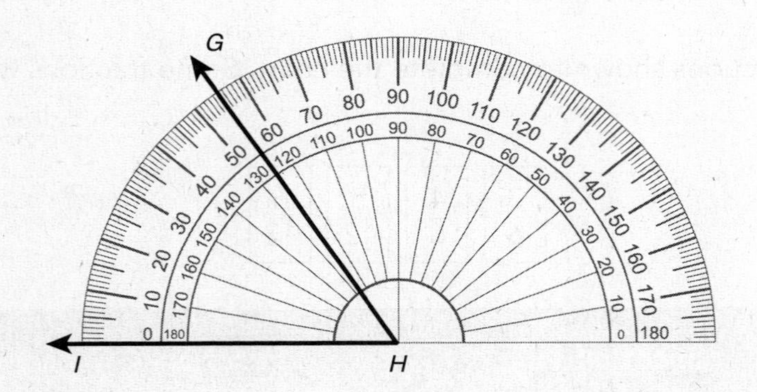

- O A. 55°
- O B. 65°
- O C. 125°
- O D. 135°

PART B

The angle is part of a shape.
Circle the number that correctly completes the statement.

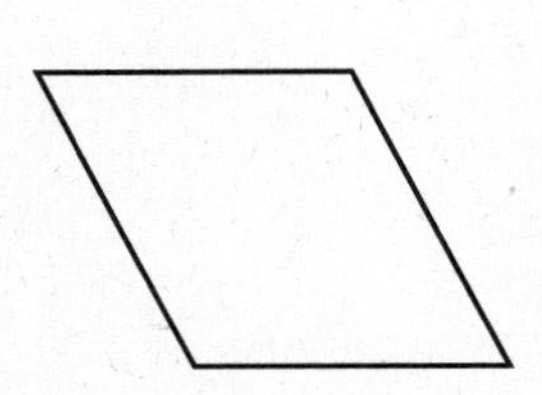

The shape has exactly _____ pair(s) of parallel sides.

1
2
4

11. A multiplication equation is shown below:

$$3 \times \frac{4}{5} = ?$$

Which has the same value? Choose TWO that are correct.

☐ A. $\frac{7}{5}$

☐ B. $12 \times \frac{1}{5}$

☐ C. $\frac{7}{8}$

☐ D. $12 \times \frac{1}{8}$

☐ E. $\frac{12}{5}$

☐ F. $7 \times \frac{1}{8}$

12. Place the decimals on the lines to correctly complete the number sentences.

11.80	11.05	11.9	11.1	11.50

11.8 < ________ 11.5 = ________

13. The diagram shows the lengths and widths of a room.

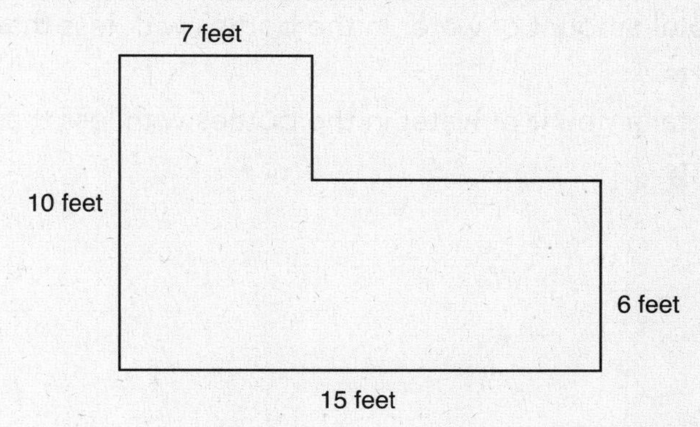

Which shows the area, in square feet, of the room?

O A. 90 square feet

O B. 98 square feet

O C. 118 square feet

O D. 150 square feet

14. Some of the girls wrote their names in capital letters. Which name has 5 letters that all have at least 1 line of symmetry?

O A. AMAYA

O B. HANNA

O C. SASHA

O D. TALIA

15. The line plot below shows the amount of water in some bottles:

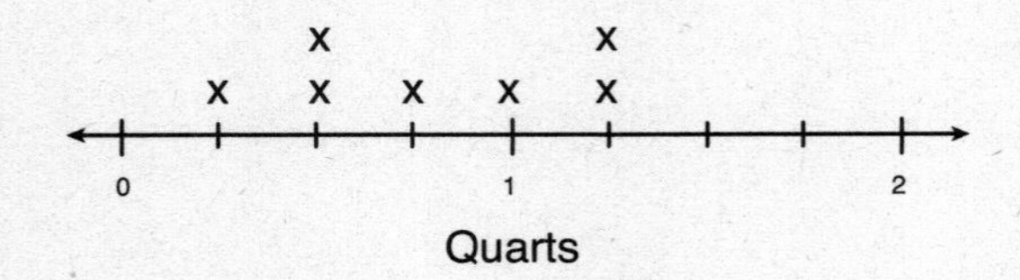

Which statements about the amounts of water are correct?
Choose THREE that are correct.

☐ A. The difference between the greatest and least amount is $1\frac{1}{4}$ quarts.

☐ B. The total amount of water in all of the bottles is $5\frac{2}{4}$ quarts.

☐ C. The difference between the greatest and least amount is 1 quart.

☐ D. The total amount of water in all of the bottles is $4\frac{2}{4}$ quarts.

☐ E. The total amount of water in the bottles with less than 1 quart is 3 quarts.

☐ F. The total amount of water in the bottles with less than 1 quart is 2 quarts.

16. The table below shows the number of inches equal to feet.

Number of Feet	3	4	6
Number of Inches	36	48	72

Circle the words and numbers to correctly complete the statement shown below:

To convert from feet to inches, ___________ the number of feet by _____.

divide — 6
subtract — 12
multiply — 24

17. Jack has 2.4 liters of ice cream in 3 equal-sized containers. How many milliliters of ice cream is in each of the 3 containers?

O A. 72 milliters
O B. 80 milliliters
O C. 720 milliliters
O D. 800 milliliters

18. Patrick bought a candle and a card to give to his mother. The price of the candle was $7.75 and the card was $2.79, both including tax. He paid for the candle and the card with $20.00. How much change did Patrick receive from the $20.00? Explain how you found your answer.

Session 2 (55 minutes)

You may NOT use a calculator for these questions. The estimated time for completing this session is 55 minutes; however, you may be permitted more time if you need it. Follow all directions to complete each question.

1. Circle the words and numbers to correctly complete the statements.

The number 8,731 is multiplied by 10.

PART A

The digit 8 in the resulting product is in the ________________ place.

hundreds
thousands
ten thousands

PART B

The digit 3 in the resulting product has a value of __________.

30
300
3,000

2. Caroline turned a door handle. The door handle moved around exactly 240 of the 1-degree angles shown.

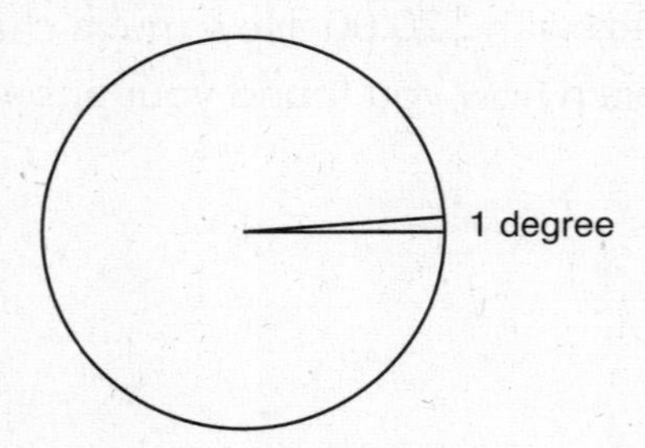

Which shows the number of degrees that the door was turned by Caroline?

O A. 1
O B. 24
O C. 100
O D. 240

3. A puppy has a weight of 8 pounds. The puppy's mother has a weight that is 17 times as much as the puppy's weight. Use the numbers shown below to write an equation that could be used to find the weight of the puppy's mother.

8	9	17	25	56	136	+	−	×	÷

☐ ☐ ☐ = ☐ pounds

4. Greta has 12 stickers.

$\frac{1}{3}$ of the stickers are flowers.

$\frac{1}{4}$ of the stickers are stars.

$\frac{5}{12}$ of the stickers are happy faces.

PART A

Greta has the most of which type of sticker? ____________

PART B

Write a fraction equal to $\frac{1}{3}$ with a denominator of 12.

$$\frac{1}{3} = \frac{?}{12}$$

5. Solve: $5 \times \square = 7{,}045$

6. The table below shows the time that Frankie spent doing his homework each night for 3 nights.

Night	Homework Time
Monday	1 hour, 7 minutes
Tuesday	76 minutes
Wednesday	70.8 minutes

What is the difference, in minutes, between the greatest and least times on the chart?

____

7. Which THREE factor pairs are the factors of 24?

☐ A. 2×12
☐ B. 3×7
☐ C. 4×6
☐ D. 5×5
☐ E. 8×3
☐ F. 9×4

8. Solve: $\square - 2{,}804 = 167$

9. In which equations does $n = 6$? Choose THREE that are correct.

☐ A. $4 + n = 2 \times 5$
☐ B. $2 + n = 24 \div 4$
☐ C. $5 \times 7 = 41 - n$
☐ D. $49 + 5 = 9 \times n$
☐ E. $60 - 11 = 10 \times n$

10. Anthony measured the length of a picture and found the perimeter.

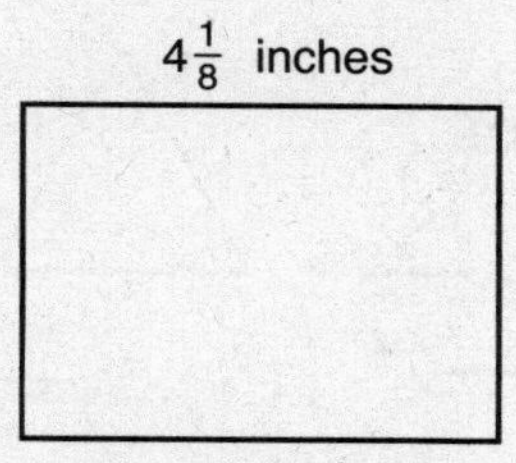

Perimeter = $14\frac{2}{8}$ inches

Which shows the width, in inches, of the picture?

O A. 3 inches

O B. 6 inches

O C. $8\frac{2}{8}$ inches

O D. $11\frac{1}{8}$ inches

11. The combined weight of 2 bags of rocks is 2.3 kilograms. Place the weights in the boxes to show a possible weight, in grams, of each of the bags.

1.0	1.1	1.2	1.3	1,000	1,100	1,200	1,300

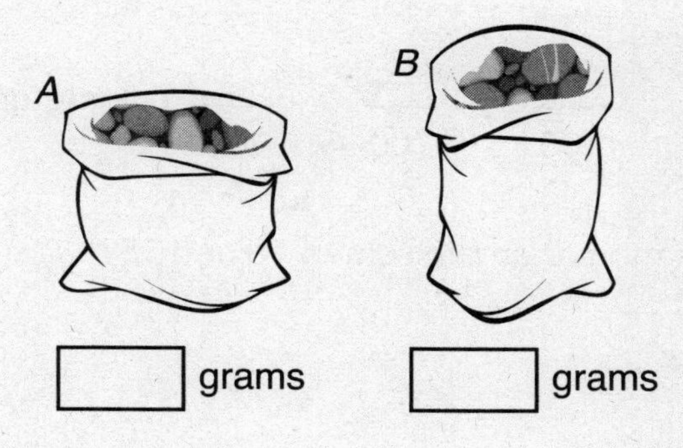

12. Joseph drew some shapes shown below:

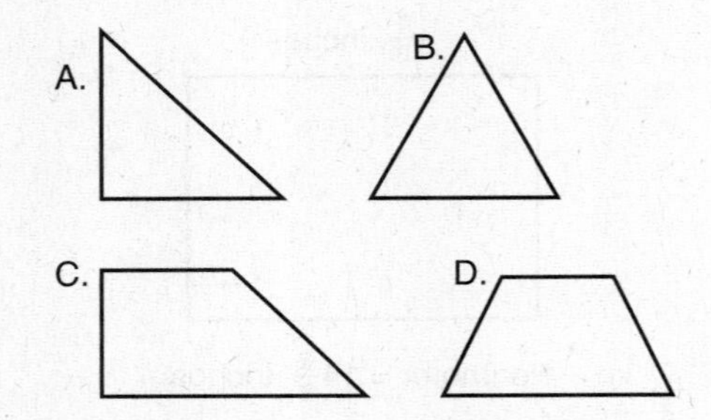

Circle Y (yes) or N (no) to correctly complete each statement.

A. Shape A is a right triangle.	Y	N
B. Shape B is a right triangle.	Y	N
C. Shape C has 2 pairs of perpendicular sides.	Y	N
D. Shape D has 2 pairs of perpendicular sides.	Y	N

13. Drew walked his dog for a distance of 0.3 kilometers in the morning and for a distance of 0.55 kilometers in the afternoon. Place the numbers in the boxes to correctly complete an equation that could be used to find the total distance, in kilometers, that Drew walked in the morning and afternoon.

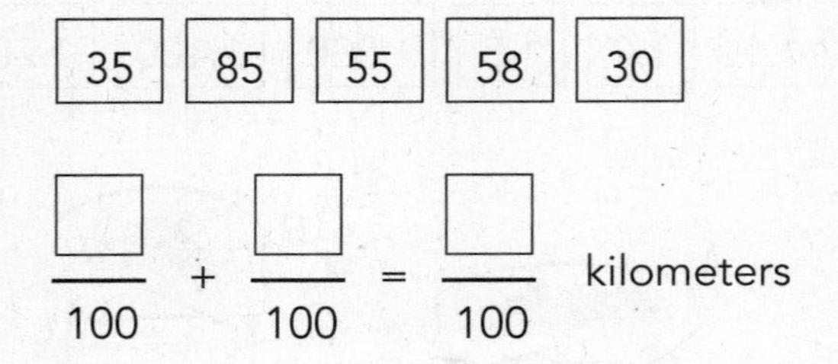

14. What is 24 times more than the sum of 18 and 15?

15. Angle *PQT* is made up of 3 smaller angles, and measures 120°.

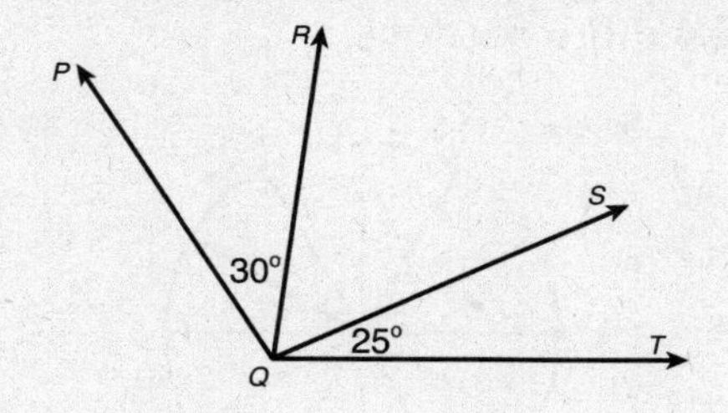

Which shows the measurement of angle *RQS*?

O A. 55°
O B. 65°
O C. 95°
O D. 145°

16. Which line is a line of symmetry?

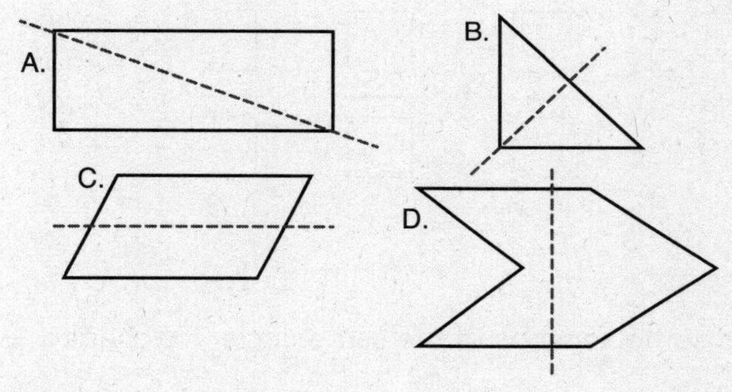

O A. Shape A
O B. Shape B
O C. Shape C
O D. Shape D

17. The circle below is cut into 12 equal sections. Adam shaded 5 of the sections and then he shaded 4 more sections.

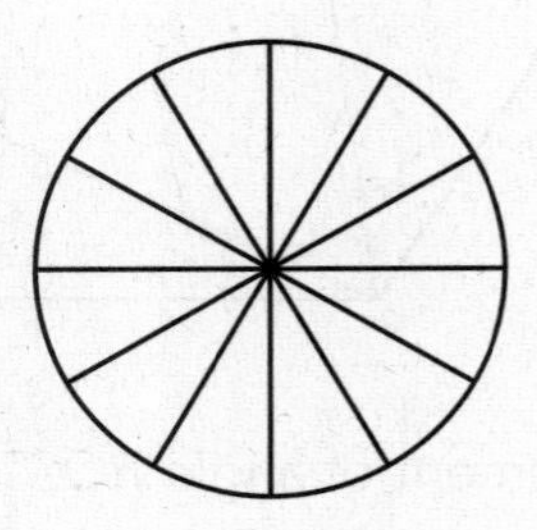

PART A

Place some of the digits in the boxes to correctly complete an equation that could be used to find the total fraction of the circle that Adam shaded. Some digits may be used more than once or not at all.

1	4	5	8	9	10	12

$$\frac{\square}{\square} + \frac{\square}{\square} = \frac{\square}{\square}$$

PART B

Josh shaded some sections of the same circle. He shaded $\frac{8}{12}$ of the circle.

Which equations could be used to find the fraction of the circle that Josh shaded? Circle Y (yes) or N (no) for each equation.

A. $4 \times \frac{2}{12} = ?$ Y N

B. $4 + \frac{4}{12} = ?$ Y N

C. $8 \times \frac{1}{12} = ?$ Y N

D. $8 + \frac{1}{12} = ?$ Y N

18. Scarlet made a table to show 2 different patterns. Both patterns start with the number 2, but have 2 different rules.

Use the numbers shown to correctly complete each number pattern.

4	6	8	10	14	16	18	22	32	64

Rule: Add 4	Rule: Multiply by 2
2, ____, ____, ____, ____, ____	2, ____, ____, ____, ____, ____

Answers to Practice Test 2

Session 1

1. **C** 8,050
   $215 \times 38 = 8,170$
   $8,170 - 120 = 8,050$

2. The number 48 is 6 times more than the number 8.
   $6 \times 8 = 48$

3. **A**, **B** $3 \times 3 = 9$, **D** $36 \div 9 = 4$

4. **C** $106,496 > 105,497$

5. Part A: Rule: – 11
   Part B: 88

6. 5,347
   $9,055 - 3,708 = 5,347$

7. 802 R4

8. $\frac{4}{6} > \frac{1}{2}$ $\left(\frac{4}{6} > \frac{3}{6}\right)$ $\frac{5}{12} < \frac{1}{2}$ $\left(\frac{5}{12} < \frac{6}{12}\right)$

9. **D** $5\frac{1}{4} - 3\frac{3}{4} = 1\frac{2}{4}$

10. Part A: **A** 55°
    Part B: The shape has exactly 2 pairs of parallel sides.

11. **B** and **E** $3 \times \frac{4}{5} = \frac{12}{5}$

12. $11.8 < 11.9$; $11.5 = 11.50$

13. **C** The shape can be cut into 2 sections.
    $(7 \times 10) + (8 \times 6) = 70 + 48 = 118$ sq ft

14. **A** AMAYA

15. **B** $\frac{1}{4} + \frac{2}{4} + \frac{2}{4} + \frac{3}{4} + 1 + 1\frac{1}{4} + 1\frac{1}{4} = 3\frac{10}{4} = 5\frac{2}{4}$
    C $1\frac{1}{4} - \frac{1}{4} = 1$
    F $\frac{1}{4} + \frac{2}{4} + \frac{2}{4} + \frac{3}{4} = \frac{8}{4} = 2$

16. To convert from feet to inches, multiply the number of feet by 12.

17. **D** 2.4 liters = 2,400 milliliters
    2,400 ÷ 3 = 800

18. Patrick received change of $9.46.
    $7.75 + $2.79 = $10.54
    $20.00 – $10.54 = $9.46

Checklist for Standards

Type and point value for each question.

Question	Standard	Type	Point Value
1	4.OA.3	I	1
2	4.OA.1	I	1
3	4.OA.4	I	1
4	4.NBT.2	I	1
5	4.OA.5	II	2
6	4.NBT.4	I	1
7	4.NBT.6	I	1
8	4.NF.2	II	2
9	4.NF.3	I	1
10	4.MD.6, 4.G.1	I	2
11	4.NF.3	I	1
12	4.NF.7	I	1
13	4.MD.3	I	1
14	4.G.3	I	1
15	4.MD.4	I	1
16	4.MD.1-2	I	1
17	4.MD.1-2	I	1
18	4.MD.1-2	II	2

Session 2

1. Part A: The digit 8 in the resulting product is in the ten thousands place.
   Part B: The digit 3 in the resulting product has a value of 300;
   $8{,}731 \times 10 = 87{,}310$
2. **D** 240
3. $8 \times 17 = 136$ pounds
4. Part A: $\frac{5}{12}$ happy faces
   Part B: $\frac{1}{3} = \frac{4}{12}$
5. $7{,}045 \div 5 = 1{,}409$
6. 9 minutes; $76 - 67 = 9$
7. **A** 2×12
   C 4×6
   E 8×3
8. 2,971 ($2{,}808 + 167 = 2{,}971$)
9. **A** $4 + 6 = 2 \times 5$
   C $5 \times 7 = 41 - 6$
   D $49 + 5 = 9 \times 6$
10. **A** 3 inches; $4\frac{1}{8} + 4\frac{1}{8} = 8\frac{2}{8}$ $14\frac{2}{8} - 8\frac{2}{8} = 6$ $6 \div 2 = 3$
11. 1,000 grams + 1,300 grams OR 1,100 grams + 1,200 grams
    2.3 kilograms = 2,300 grams
12. A. Y B. N C. Y D. N
13. $\frac{30}{100} + \frac{55}{100} = \frac{85}{100}$ $0.3 = 0.30 = \frac{30}{100}$ $0.55 = \frac{55}{100}$
14. **792** $18 + 15 = 33$ $33 \times 24 = 792$
15. **B** $30^\circ + 25^\circ = 55^\circ$ $120^\circ - 55^\circ = 65^\circ$
16. **B** Shape B

17. Part A: $\frac{5}{12}+\frac{4}{12}=\frac{9}{12}$

Part B: A. $4 \times \frac{2}{12} = \frac{8}{12}$ Y

B. $4 + \frac{4}{12} = 4\frac{4}{12}$ N

C. $8 \times \frac{1}{12} = \frac{8}{12}$ Y

D. $8 + \frac{1}{12} = 8\frac{1}{12}$ N

18. Rule: Add 4: 2, 6, 10, 14, 18, 22
Rule: Multiply by 2: 2, 4, 8, 16, 32, 64

Checklist for Standards

Type and point value for each question.

Question	Standard	Type	Point Value
1	4.OA.3	I	1
2	4.OA.1	I	1
3	4.OA.4	I	1
4	4.NBT.2	I	1
5	4.OA.5	II	2
6	4.NBT.4	I	1
7	4.NBT.6	I	1
8	4.NF.2	II	2
9	4.NF.3	I	1
10	4.MD.6, 4.G.1	I	2
11	4.NF.3	I	1
12	4.NF.7	I	1
13	4.MD.3	I	1
14	4.G.3	I	1
15	4.MD.4	I	1
16	4.MD.1-2	I	1
17	4.MD.1-2	I	1
18	4.MD.1-2	II	2

Common Core Standards, Mathematics Grade 4

Domain: Operations & Algebraic Thinking
Cluster: Use the Four Operations with Whole Numbers to Solve Problems.
Standard: *CCSS.MATH.CONTENT.4.OA.A.1* Interpret a multiplication equation as a comparison, e.g., interpret 35 = 5 × 7 as a statement that 35 is 5 times as many as 7 and 7 times as many as 5. Represent verbal statements of multiplicative comparisons as multiplication equations.
Standard: *CCSS.MATH.CONTENT.4.OA.A.2* Multiply or divide to solve word problems involving multiplicative comparison, e.g., by using drawings and equations with a symbol for the unknown number to represent the problem, distinguishing multiplicative comparison from additive comparison.
Standard: *CCSS.MATH.CONTENT.4.OA.A.3* Solve multistep word problems posed with whole numbers and having whole-number answers using the four operations, including problems in which remainders must be interpreted. Represent these problems using equations with a letter standing for the unknown quantity. Assess the reasonableness of answers using mental computation and estimation strategies including rounding.
Cluster: Gain Familiarity with Factors and Multiples.
Standard: *CCSS.MATH.CONTENT.4.OA.B.4* Find all factor pairs for a whole number in the range 1–100. Recognize that a whole number is a multiple of each of its factors. Determine whether a given whole number in the range 1–100 is a multiple of a given one-digit number. Determine whether a given whole number in the range 1–100 is prime or composite.
Cluster: Generate and Analyze Patterns.
Standard: *CCSS.MATH.CONTENT.4.OA.C.5* Generate a number or shape pattern that follows a given rule. Identify apparent features of the pattern that were not explicit in the rule itself. For example, given the rule "Add 3" and the starting number 1, generate terms in the resulting sequence and observe that the terms appear to alternate between odd and even numbers. Explain informally why the numbers will continue to alternate in this way.

Domain: Number & Operations in Base Ten
Cluster: Generalize Place Value Understanding for Multidigit Whole Numbers.
Standard: *CCSS.MATH.CONTENT.4.NBT.A.1*   Recognize that in a multidigit whole number, a digit in one place represents ten times what it represents in the place to its right. For example, recognize that $700 \div 70 = 10$ by applying concepts of place value and division.
Standard: *CCSS.MATH.CONTENT.4.NBT.A.2*   Read and write multidigit whole numbers using base-ten numerals, number names, and expanded form. Compare two multidigit numbers based on meanings of the digits in each place, using $>$, $=$, and $<$ symbols to record the results of comparisons.
Standard: *CCSS.MATH.CONTENT.4.NBT.A.3*   Use place value understanding to round multidigit whole numbers to any place.
Cluster: Use Place Value Understanding and Properties of Operations to Perform Multidigit Arithmetic.
Standard: *CCSS.MATH.CONTENT.4.NBT.B.4*   Fluently add and subtract multidigit whole numbers using the standard algorithm.
Standard: *CCSS.MATH.CONTENT.4.NBT.B.5*   Multiply a whole number of up to four digits by a one-digit whole number, and multiply two two-digit numbers, using strategies based on place value and the properties of operations. Illustrate and explain the calculation by using equations, rectangular arrays, and/or area models.
Standard: *CCSS.MATH.CONTENT.4.NBT.B.6*   Find whole-number quotients and remainders with up to four-digit dividends and one-digit divisors, using strategies based on place value, the properties of operations, and/or the relationship between multiplication and division. Illustrate and explain the calculation by using equations, rectangular arrays, and/or area models.
Domain: Number & Operations—Fractions
Cluster: Extend Understanding of Fraction Equivalence and Ordering.
Standard: *CCSS.MATH.CONTENT.4.NF.A.1*   Explain why a fraction a/b is equivalent to a fraction $(n \times a)/(n \times b)$ by using visual fraction models, with attention to how the number and size of the parts differ even though the two fractions themselves are the same size. Use this principle to recognize and generate equivalent fractions.

Standard: CCSS.MATH.CONTENT.4.NF.A.2
Compare two fractions with different numerators and different denominators, e.g., by creating common denominators or numerators, or by comparing to a benchmark fraction such as 1/2. Recognize that comparisons are valid only when the two fractions refer to the same whole. Record the results of comparisons with symbols >, =, or <, and justify the conclusions, e.g., by using a visual fraction model.

Cluster: Build Fractions from Unit Fractions.

Standard: CCSS.MATH.CONTENT.4.NF.B.3
Understand a fraction a/b with $a > 1$ as a sum of fractions $1/b$.

Standard: CCSS.MATH.CONTENT.4.NF.B.3.A
Understand addition and subtraction of fractions as joining and separating parts referring to the same whole.

Standard: CCSS.MATH.CONTENT.4.NF.B.3.B
Decompose a fraction into a sum of fractions with the same denominator in more than one way, recording each decomposition by an equation. Justify decompositions, e.g., by using a visual fraction model. Examples: $3/8 = 1/8 + 1/8 + 1/8$; $3/8 = 1/8 + 2/8$; $2\ 1/8 = 1 + 1 + 1/8 = 8/8 + 8/8 + 1/8$.

Standard: CCSS.MATH.CONTENT.4.NF.B.3.C
Add and subtract mixed numbers with like denominators, e.g., by replacing each mixed number with an equivalent fraction, and/or by using properties of operations and the relationship between addition and subtraction.

Standard: CCSS.MATH.CONTENT.4.NF.B.3.D
Solve word problems involving addition and subtraction of fractions referring to the same whole and having like denominators, e.g., by using visual fraction models and equations to represent the problem.

Standard: CCSS.MATH.CONTENT.4.NF.B.4
Apply and extend previous understandings of multiplication to multiply a fraction by a whole number.

Standard: CCSS.MATH.CONTENT.4.NF.B.4.A
Understand a fraction a/b as a multiple of $1/b$. For example, use a visual fraction model to represent 5/4 as the product $5 \times (1/4)$, recording the conclusion by the equation $5/4 = 5 \times (1/4)$.

Standard: CCSS.MATH.CONTENT.4.NF.B.4.B
Understand a multiple of a/b as a multiple of $1/b$, and use this understanding to multiply a fraction by a whole number. For example, use a visual fraction model to express $3 \times (2/5)$ as $6 \times (1/5)$, recognizing this product as 6/5. (In general, $n \times (a/b) = (n \times a)/b$.)

Standard: CCSS.MATH.CONTENT.4.NF.B.4.C
Solve word problems involving multiplication of a fraction by a whole number, e.g., by using visual fraction models and equations to represent the problem. For example, if each person at a party will eat 3/8 of a pound of roast beef, and there will be 5 people at the party, how many pounds of roast beef will be needed? Between what two whole numbers does your answer lie?

Cluster: Understand Decimal Notation for Fractions, and Compare Decimal Fractions.

Standard: CCSS.MATH.CONTENT.4.NF.C.5
Express a fraction with denominator 10 as an equivalent fraction with denominator 100, and use this technique to add two fractions with respective denominators 10 and 100. For example, express 3/10 as 30/100, and add 3/10 + 4/100 = 34/100.

Standard: CCSS.MATH.CONTENT.4.NF.C.6
Use decimal notation for fractions with denominators 10 or 100. For example, rewrite 0.62 as 62/100; describe a length as 0.62 meters; locate 0.62 on a number line diagram.

Standard: CCSS.MATH.CONTENT.4.NF.C.7
Compare two decimals to hundredths by reasoning about their size. Recognize that comparisons are valid only when the two decimals refer to the same whole. Record the results of comparisons with the symbols >, =, or <, and justify the conclusions, e.g., by using a visual model.

Domain: Measurement & Data

Cluster: Solve Problems Involving Measurement and Conversion of Measurements.

Standard: CCSS.MATH.CONTENT.4.MD.A.1
Know relative sizes of measurement units within one system of units including km, m, cm; kg, g; lb, oz.; l, ml; hr, min, sec. Within a single system of measurement, express measurements in a larger unit in terms of a smaller unit. Record measurement equivalents in a two-column table. For example, know that 1 ft is 12 times as long as 1 in. Express the length of a 4 ft snake as 48 in. Generate a conversion table for feet and inches listing the number pairs (1, 12), (2, 24), (3, 36), ...

Standard: CCSS.MATH.CONTENT.4.MD.A.2
Use the four operations to solve word problems involving distances, intervals of time, liquid volumes, masses of objects, and money, including problems involving simple fractions or decimals, and problems that require expressing measurements given in a larger unit in terms of a smaller unit. Represent measurement quantities using diagrams such as number line diagrams that feature a measurement scale.

Standard: CCSS.MATH.CONTENT.4.MD.A.3

Apply the area and perimeter formulas for rectangles in real world and mathematical problems. For example, find the width of a rectangular room given the area of the flooring and the length, by viewing the area formula as a multiplication equation with an unknown factor.

Cluster: Represent and Interpret Data.

Standard: CCSS.MATH.CONTENT.4.MD.B.4

Make a line plot to display a data set of measurements in fractions of a unit (1/2, 1/4, 1/8). Solve problems involving addition and subtraction of fractions by using information presented in line plots. For example, from a line plot find and interpret the difference in length between the longest and shortest specimens in an insect collection.

Cluster: Geometric Measurement: Understand Concepts of Angle and Measure Angles.

Standard: CCSS.MATH.CONTENT.4.MD.C.5

Recognize angles as geometric shpes that are formed wherever two rays share a common endpoint, and understand concepts of angle measurement.

Standard: CCSS.MATH.CONTENT.4.MD.C.5.A

An angle is measured with reference to a circle with its center at the common endpoint of the rays, by considering the fraction of the circular arc between the points where the two rays intersect the circle. An angle that turns through 1/360 of a circle is called a "one-degree angle," and can be used to measure angles.

Standard: CCSS.MATH.CONTENT.4.MD.C.5.B

An angle that turns through n one-degree angles is said to have an angle measure of n degrees.

Standard: CCSS.MATH.CONTENT.4.MD.C.6

Measure angles in whole-number degrees using a protractor. Sketch angles of specified measure.

Standard: CCSS.MATH.CONTENT.4.MD.C.7

Recognize angle measure as additive. When an angle is decomposed into non-overlapping parts, the angle measure of the whole is the sum of the angle measures of the parts. Solve addition and subtraction problems to find unknown angles on a diagram in real world and mathematical problems, e.g., by using an equation with a symbol for the unknown angle measure.

Domain: Geometry
Cluster: Draw and Identify Lines and Angles, and Classify Shapes by Properties of Their Lines and Angles.
Standard: CCSS.MATH.CONTENT.4.G.A.1 Draw points, lines, line segments, rays, angles (right, acute, obtuse), and perpendicular and parallel lines. Identify these in two-dimensional figures.
Standard: CCSS.MATH.CONTENT.4.G.A.2 Classify two-dimensional figures based on the presence or absence of parallel or perpendicular lines, or the presence or absence of angles of a specified size. Recognize right triangles as a category, and identify right triangles.
Standard: CCSS.MATH.CONTENT.4.G.A.3 Recognize a line of symmetry for a two-dimensional figure as a line across the figure such that the figure can be folded along the line into matching parts. Identify line-symmetric figures and draw lines of symmetry.

Index